Routledge R

The Modern Urban Landscape

First published in 1987, this book provides a wide-ranging account of how modern cities have come to look as they do — differing radically from their predecessors in scale, style, detail and meaning. It uses many illustrations and examples to explore the origins and development of specific landscape features. More generally it traces the interconnected changes which have occurred in architectural and aesthetic fashions, in economic and social conditions, and in planning — which together have created the landscape that now prevails in most of the cities of the world. This book will be of interest to students of architecture, urban studies and geography.

The Modern Urban Landscape

Edward Relph

Routledge
Taylor & Francis Group

First published in 1987
by Croom Helm

This edition first published in 2016 by Routledge
2 Park Square, Milton Park, Abingdon, Oxon, OX14 4RN
and by Routledge
711 Third Avenue, New York, NY 10017

Routledge is an imprint of the Taylor & Francis Group, an informa business

© 1987 E. Relph

Publisher's Note
The publisher has gone to great lengths to ensure the quality of this reprint but points out that some imperfections in the original copies may be apparent.

Disclaimer
The publisher has made every effort to trace copyright holders and welcomes correspondence from those they have been unable to contact.

A Library of Congress record exists under LC control number: 87003809

ISBN 13: 978-1-138-66769-3 (hbk)
ISBN 13: 978-1-315-61886-9 (ebk)
ISBN 13: 978-1-138-66775-4 (pbk)

THE MODERN URBAN LANDSCAPE

EDWARD RELPH

CROOM HELM
London & Sydney

© 1987 E. Relph

Croom Helm Ltd, Provident House, Burrell Row,
Beckenham, Kent, BR3 1AT

Croom Helm Australia, 44-50 Waterloo Road,
North Ryde, 2113, New South Wales

British Library Cataloguing in Publication Data

Relph, E.
 The modern urban landscape: 1880 to the
 present.
 1. Landscape architecture—History—19th
 century   2. Landscape architecture—
 History—20th century   3. City planning—
 History
 I. Title
 712'.09'04   SB472

ISBN 0-7099-2231-0
ISBN 0-7099-4270-2 Pbk

Typeset in 10pt Baskerville Roman by
Leaper & Gard Ltd, Bristol, England
Printed and bound in Great Britain by
Biddles Ltd, Guildford and King's Lynn

Contents

Preface

The late twentieth century may be the first period in history when it is possible for most people to survive without first-hand knowledge of their surroundings. It is now quite possible to get around a city by using borrowed information, reading guide books and following signs. I find this depressing because the landscapes and places we live in are important. Whether we shape them or they shape us, they are expressions of what we are like. Our lives are impoverished precisely to the extent that we ignore them.

Like the clothes we wear, landscapes not only hide but also reveal a great deal about what lies beneath and within. My hope is that *The modern urban landscape* will encourage its readers to look for themselves with fresh eyes at townscapes and cityscapes, and that it will provide a foundation for making sense of what they have noticed. The focus of this book is on the landscapes of large cities because it is in the streets and buildings of these that the effects of the present age are most concentrated and most obvious. In smaller cities and towns what is new is usually appended to the edges or interspersed with the old, and is therefore less overwhelming, though it is still significant as the distinctive contribution of the twentieth century to urban landscapes. The approach is to consider how modern cities and the new parts of towns have come to look as they do by tracing the separate yet interconnected changes which have occurred in architecture, planning, technology and social conditions since about 1880. So this book is in part a review of familiar historical developments, such as the rise of town planning and modernist architectural styles. But it also adds to these, casts them in a new light, and puts them in context by interpreting them in terms of their contribution to the overall appearance of cities.

It would have been impossible for me to have completed a broad-ranging study such as this one without help from colleagues and students. I want to acknowledge the assistance of the following in suggesting sources, lending books, offering comments, inviting me to conferences in exotic places, providing travel assistance, taking me on field excursions, undertaking bits of research, and otherwise contributing, albeit unwittingly, to the writing of this book: Kim Dovey, David Seamon, Randy Hester,

Marcia McNally, Paul Groth, Peirce Lewis, Michael Bunce, Richard Harris, Jim Lemon, Deryck Holdsworth, Gunter Gad, Shoukry Roweis, Ed Jackson, Barry Murphy, Patsy Eubanks, Helen Armstrong, Chris Maher, Lucy Jarosz, Hong-Key Yoon, Peter Perry, Seamus Smythe, Madis Pihlak, John Punter, Rick Peddie, Suzanne Mackenzie, Len Guelke, Neil Evernden, Rod Watson, Michele Bouchier, Miriam Wyman, Chuck Geiger, Dick Morino, Katy Oliver, Jane Bonshek, Robin Kearns, Mary Marum, Wayne Reeves, Evelyn Ruppert, Craig Hunter, Derek Dalgleish, Brian Banks, Ross Nelson, and Nigel Hall. My thanks to Liz Lew for her help with the photographs; and to the Social Science and Humanities Research Council of Canada, the University of Toronto Office of Research Administration, the Council for Education in Landscape Architecture, the Canadian Society of Landscape Architects, the College of Environmental Design at the University of California at Berkeley, the University of Alberta, the Department of Architecture at the Royal Melbourne Institute of Technology, PAPER, the Department of Geography at Rutgers University, and Oberlin College.

I am especially grateful to David Harford of the Graphics Department at the Scarborough Campus of the University of Toronto for preparing the illustrations from a mountain of slides, negatives and books, and to Peter Sowden of Croom Helm and George Thompson of The Johns Hopkins Press for their editorial wisdom and encouragement.

Irene, Gwyn and Lexy suffered through numerous side-trips to places they were not much interested in and will no doubt be glad to see the whole thing over and done with. They always help me to maintain a sense of perspective.

Highland Creek

1

Introduction

A century ago there were no skyscraper offices, no automobile-dependent suburbs, no streets bathed at night in the glare of electric lights, no airports, parking lots, expressways or shopping malls; there were no microwave transmission towers or huge concrete convention centres or international chains of fast food restaurants. These, and most of the other familiar features of modern cities, had yet to be invented and built. Slowly at first, then increasingly rapidly in the second half of the twentieth century, they have been put together to create an urban landscape that bears little resemblance to any of its industrial, renaissance or medieval predecessors, even though it has sometimes been built on their footprint of street and lot patterns. For anyone living in a city this new landscape is omnipresent, and even for those who live in quaint old towns and remote hamlets it is an unavoidable, encroaching reality. It is encountered wherever we see skyscraper skylines, electrical signs, concrete buildings or parking lots, and whenever we eat an international hamburger in the climate-controlled gloss of an indoor shopping centre, or look up at the cloud reflections in the mirror glass of an office building, or suffer the multiple indignities of the frantic spaces of an airport, or, most frequently of all, whenever we drive a car.

In spite of the familiarity and virtual omnipresence of modern urban landscapes they must be generally seen as unremarkable or unpleasant because nobody pays much serious attention to them. It is almost as though they have been designed not to be noticed. There are thousands of books, both academic and interesting, which examine the structure and form of twentieth-century cities; of these only a handful has anything to say about their appearance. And of this handful most treat whatever is new with disdain or with shrill condemnation. Poets and painters almost completely ignore it. Even W.G. Hoskins, in his excellent book *The making of the English landscape* (1959), could not bring himself to consider anything modern because he found that every single change in the English landscape since the later years of the nineteenth century had 'either uglified it or destroyed its meaning, or both' (p. 298).

This is hardly fair. Modern landscapes deserve to be understood and appreciated. The changes that have been effected over the last century have, both in character and scale, been simply enormous. Like them or not, for the great majority of us they are the context of daily life and therefore merit at least some small part of our critical attention. They are also, by almost any standard, one of the great constructive accomplishments of the modern age. It is difficult to be precise about such things, but population increases alone suggest that about 60 per cent of the population of the developed world must live and work in places made since 1945. The making of these has required substantial investments of money, time and effort, so it is safe to assume that their appearance is neither accidental nor incidental. Furthermore, the sheer scale and durability of the materials of modern landscapes guarantee they will be a major part of the legacy of this age to the future, one that will directly inform our descendants about the values and abilities of twentieth-century society, just as gothic cathedrals and medieval townscapes tell us something of the world of the Middle Ages. Perhaps in 500 years time tourists to heritage districts at the World Trade Center in Manhattan or the Barbican in London will gaze in awe at these huge towers, vacuous spaces and concrete palisades, and marvel that there could ever have been a society capable of creating such places. This prospect alone seems to me to be sufficient cause for an enquiry into the development of modern urban landscapes and the values which they enfold and express.

The aim of this book is to give an account of the development of the appearance of cities over the last 100 years in order to explain how they have come to look as they do. Few influences on modern landscapes can be traced back before the technological and social changes which occurred in the 1880s, so I confine my attention to the period from then to the present. There have, of course, also been dramatic changes to the appearance of the countryside in this period, but it is in cities that modern development has been most concentrated and it is the look of cities that commands my attention.

Landscapes are the visual contexts of daily existence, though I do not suppose many people often use the actual word 'landscape' to describe what they see as they walk down the street or stare through the windshield of a car. Nevertheless we manipulate landscapes in gardens, take quiet pleasure in seeing them outlined against a sunset or highlighted when the sun comes out after a thunderstorm, we consume them as tourists and record them thoughtlessly on film (Figure 1.1). They are easily photographed — merely point a camera out of doors (or even indoors in the enclosed landscape of a shopping mall or atrium, though taking photographs of these does tend to annoy security guards). All of which suggests that landscapes are obvious things. Yet when we try to analyse them it soon turns out, first of all, they are so familiar and all-embracing it is hard to get them into a clear perspective, and then that they cannot be easily disassembled into their component parts, such as buildings and roads, without losing a sense of the whole scene. So landscapes are at once obvious and elusive; it seems we know exactly what they are until we try to think and write about them, or to change them in some way, and then they become enigmatic and fragile.

Unfortunately the easiest way of coping with this fragile wholeness is to ignore it and to deal with fragments of landscape as though they are isolated from a context. There are, for example, often signs at construction sites with paintings (usually called 'Artist's conception') to show what the finished project will look like, yet which depict it standing in splendid isolation with the surroundings blanked out (Figure 1.2). No matter how hard I try I seem to be unable to affect this sort of blinkered vision, to concentrate as an architectural historian might on one building, to isolate planning or social conditions or some other single factor as the primary agent of landscape change. My unspecialised understanding is that buildings always have a context, that they are visibly related to the spaces and structures around them, and to the planned layouts and forms of streets, and that they are the products of technological developments and social circumstances. To try to understand landscapes or cities by isolating one aspect for detailed attention is like attempting to describe an entire human being on the basis of a detailed study of their feet or their cooking. Serious misrepresentations are bound to be involved.

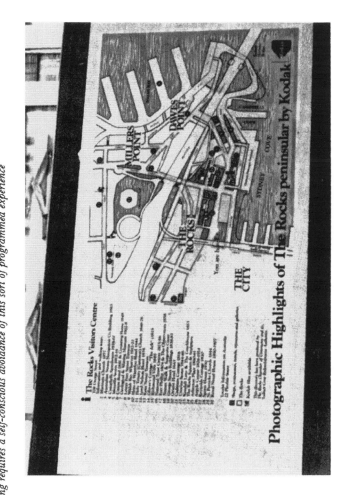

Figure 1.1: A prescription for seeing: a sign at The Rocks in Sydney informs visitors exactly where to take photographs (and to buy film). Effective landscape watching requires a self-conscious avoidance of this sort of programmed experience

I prefer to start with the totality of what I see, and to try to puzzle out its appearance by following several directions more or less at once. This approach, which I suppose could be considered geographical, puts general understanding ahead of specialised explanation. It is, quite frankly, not fashionable in this age of expertise. Yet I cannot help but believe that the generalist's view of reality is as valid as that seen through a microscope or an econometric model, and it is certainly a good deal more accessible and vital. Specialised studies may offer depth, but general accounts such as this one provide broad outlines and context. An important implication of this broad approach to landscape is that, while books about architecture, planning, technological developments and social history have valuable details, the best sources of information about landscape are landscapes themselves. Accordingly I have spent several years looking closely at the modern parts of towns and cities I have visited, most of them in North America and Britain, some in Europe and Australia and New Zealand; I have attended both to their obvious forms and their enigmatic features, and I have wondered about what they reveal of the people who chose to give them this appearance.

Someone, and I regret I cannot remember who since the term seems so appropriate, calls this approach simply 'watching', watching for unusual details, for new developments, for insights into what lies behind recent fashions, for ironic juxtapositions of signs (one of my favourites is the street name Production Drive mounted above the information sign — Dead End). I have sketched fragments of plazas, explored outside stairways that apparently lead nowhere (there are a surprising number of these — probably something to do with fire regulations), I have taken photographs of city skylines while driving on expressways, and I have walked and watched where most people are content only to drive, though I have tried not to impose a pedestrian's perspective where the driver's landscape prevails. In my watching I have come upon fascinating things which reveal human foibles and accomplishments, such as the northern Pennsylvanian custom of decorating the front yard at Easter with plastic inflatable rabbits, and complex self-help housing projects in South Wales. I have found there are no modern landscapes which are not informative and interesting, though, paradoxically and sadly given the time and money invested, there are fewer and fewer which seem to promote any sorts of pleasures or personal freedoms except those

Figure 1.2: The architect's view of 'the building in splendid isolation': Seventh Avenue just north of Times Square, New York

which are based in efficiency of function and material satisfactions.

Just as it is not possible for an artist, even a magic realist, to paint every blade of grass and every brick, so it has not been possible for me to consider every aspect of modern landscapes. My attention has been directed mostly to the widely seen forms of the built environments of cities — their structures, their streets and their spaces. There are many things that I have not been able to include, for instance there is no detailed consideration of the ways people personalise their properties, nor of the subterranean landscape of subway systems and pedestrian tunnels which is so much a part of the daily experience of citizens of London, Paris, New York, Toronto and many other metropolitan areas. And, because my main concern is to describe the broad outlines of the development of modern urban landscapes, I pay relatively little attention to regional and national differences. In spite of all the forces encouraging standardisation such differences do persist, for example because of international urban design fashions of the 1980s which paradoxically encourage local architectural styles. In Quebec many suburban houses have traditional bell-cast roofs, and in Australia an outback vernacular with galvanised iron details is widely used for new buildings. Even at the height of the popularity of international modernism there were regional variations in architecture and streetscape, variations which resulted from locally developed design standards, the availability and cost of materials and so on. Like Coca-Cola advertisements around the world, public housing projects in America, in Britain and in the Netherlands are not identical. They do, however, have profound similarities because they all derive from a basic pattern which was conceived so that it would be appropriate anywhere and could be used internationally with little regard for regional tradition.

As I put together my observations with information from documentary sources it became clear that there are four particularly important and interrelated influences which account both for the historical distinctiveness and for the similarities in appearance of late twentieth-century cities. These are — in no special order — architecture, technological innovations, planning, and social developments. Buildings, whether architect designed, hand-made or mass-produced, are the most obvious human artefacts in urban landscapes, and it takes little looking at modern architecture to realise that its unornamented angular forms are governed by aesthetic principles quite different from those which prevailed in earlier centuries. Technological advances of the last 100 years, such as structural steel, commercial electricity and automobiles, are significant because they have made possible entirely new built-forms and ways of life. Urban planning was, for most modern intents and purposes, invented at the beginning of the present century, partly to protect us from our own worst tendency to exploit others and partly to realise utopian visions of cities in which good health, justice and equality prevailed; in practice it has come to determine the layouts and arrangements of almost all the elements of cities. And modern urban societies of the developed world differ from their antecedents because they are mostly literate, comfortably housed, healthy, engaged in sedentary brain-work (or rationalised unemployment), and bonded together by systems of instantaneous electronic communication rather than kinship; these sorts of social changes have left clearly visible results in the consumer landscapes of the suburbs and the corporate landscapes of skyscraper offices.

Landscapes are substantial if intangible things. They should not be thought of as mirrors which happen to catch the otherwise invisible image of the spirit of the times. They are, however, made within a context of well-attested ideas and beliefs about how the world works, and how it might be improved. There are two ideas which have had an especially important impact on urban landscapes during the last 100 years.

The first of these is internationalism. Although ideas and fashions have never been completely confined by national boundaries their dispersal was previously limited by slowness of travel, so that they were usually adapted to existing regional traditions as they were taken from one part of the globe to another. The result was a marked regional variety based in local customs and architecture, the sort of thing that can still be seen in remnants of old landscapes in France or England. During the last century new building technologies and faster communications have greatly reduced the possibilities for regional adaptations. Contemporaneously the inclination of many leading businessmen, architects and planners has been to travel widely, bearing and borrowing designs which would work equally well everywhere. For those for whom travel was too expensive an entire bevy of professional journals, including *Town and Country Planning*, *The American City* and *The Architectural Record*, began publication between about 1890 and 1910 and provided vicarious knowledge of the international developments of the day. They have been joined subsequently by countless others. All of this has usually been judged to be a good thing, and commercial products, architectural fashions and planning practices have been increasingly exported, imported, borrowed, copied and deliberately designed for international consumption. The result is that virtually identical new bits of cities now seem to crop up almost everywhere, and behind any national or regional differences that might be visible there are always widely shared patterns and an international habit of thought.

A less immediately apparent idea that pervades twentieth-century thinking is a conviction in the merits of selfconsciousness. It seems that everything now is subjected to cool analysis and technical manipulation, leaving little room for the traditions which stood behind most pre-industrial landscapes. Even commonplace objects, such as curbstones and parking meters, have been deliberately invented to solve specific problems, given

Figure 1.3: The Barbican Centre in London, no less than a Greek temple or a medieval cathedral, informs us about contemporary standards of truth, beauty and goodness

particular forms on a drafting table, examined by committees, made the object of design standards, and then installed and maintained by teams of specially trained workmen. Similar design approaches are used for buildings, neighbourhoods, social plans, and entire new towns. As a result the modern urban landscape is both rationalised and artificial, which is another way of saying that it is intensely human, an expression of human will and deeply imbued with meaning, though it is sometimes hard to remember this (Figure 1.3). A major modern development such as the Barbican in London, no less than the temples of Greece or the cathedrals of medieval England, informs us about prevailing standards of truth, beauty and goodness.

2

Looking Back at the Future: Late Twentieth-century Landscapes in the 1890s

Landscapes are made by ideas as well as by construction, and the last decade of the nineteenth century was filled with ideas about the ways societies and cities might be in the future. Hopes ran high. A flurry of technological innovations promised dramatic improvements in everyday comfort and health, while the still fresh philosophies of socialism suggested that these improvements would soon be made equally available to everyone. Utopian speculation flourished as a serious effort to determine the character of the glorious future that seemed to be emerging.

In 1888 Edward Bellamy published *Looking backward*, an account of life in Boston in the year 2000. His book caught the popular imagination, sold several million copies, was translated into 20 languages, and led to the founding of many local societies to promote his ideas. The response was a little less dramatic to *News from nowhere*, which William Morris published in 1890 as a rebuttal to Bellamy's vision of the future. Nevertheless, Morris was already well known for his advocacy of craftsmanship and was also a central figure in the English socialist movement, so his ideas about life in England in the early twenty-first century were taken seriously.

These two books mark the inception of modern, future-oriented thinking about cities and landscapes. Indeed, delight at the prospects of the future has never since been so sweet, and has come to be replaced by a sense of deep apprehension. If we think ahead now it is cautiously, to the end of the century, and with as much attention to what might be worse as to what might be better. In the 1890s there was none of this doubt. 'The Golden Age lies before us and not behind us,' Bellamy wrote (p. 222), 'and is not far away.' This was refreshingly new. For centuries people had looked to the past for inspiration, a habit which had most recently manifest itself in a series of unimaginative or confused revivals of gothic and classical architecture. Now they turned to the future with a confidence that social and economic problems could be solved by the simple method of remaking the places in which people lived. Bellamy and Morris were at the forefront of this radical reorientation. Previous utopian writers had mostly set their ideal societies in remote corners of the world, but geographical exploration and utopian disappointment went hand in hand; by the 1880s almost all the globe had been explored and no utopias had been discovered. So Bellamy and Morris were left with little choice except to imagine future societies in which the poverty, grime and inequalities of their own age had been overcome, and by doing this they gave form to the dramatic shifts in political thought and technology that were then occurring. It was, of course, social and economic matters that concerned them most, and their ideal future landscapes were little more than imaginative packaging, or expressions of personal taste and whimsy. Nevertheless, the differences between the late-Victorian urban landscapes which they so abhorred, the future ones they imagined, and what exists now, are valuable measures of the originality and unexpectedness of all that has happened in between.

Victorian inequalities

In the opening pages of *Looking backward* Bellamy presents a compelling image of late nineteenth-century society. It was, he wrote, as though the poor masses of humanity were harnessed to a coach which they had to drag along a very hilly and sandy road. On the coach, on comfortable seats from which they never descended, were the prosperous members of society, enjoying the scenery and discussing the merits of the team pulling them along. These passengers felt great compassion for those labouring on their behalf, and often called down words of sympathy and encouragement, yet at the same time they did everything in their power to avoid being jolted out of their secure seats and having to join the masses.

This image is hardly overdrawn. In *Crown of wild olive* (1865) John Ruskin quotes a report from the Telegraph of 16 January 1865 describing 'the paupers of the Andover Union gnawing the scraps of putrid flesh and sucking the marrow from the bones of horses which they were employed to crush' (p. 45). At the other social extreme was Louis Tiffany, who in the 1890s had a house built for himself in New York City with imported Sudanese village huts incorporated into the landings of the grand staircase, so that his guests might rest in them as they ascended to the banqueting hall for a 14-course meal. On the one hand — conspicuous consumption (the phrase was coined by Thorstein Veblen at the turn of the century to describe just the sort of ostentation that Tiffany displayed); on the other hand — conspicuous poverty. It was the poverty which was particularly offensive to reformers and utopian writers. In a flashback scene Edward Bellamy imagined himself transported from the year 2000 back to the Boston of 1887:

> I found myself in the midst of a scene of squalor and human degradation such as only the South Cove tenement district could present. I had seen the mad wasting of human labour; here I saw in direct shape the want that waste had bred.
>
> From the black doorways and windows of the rookeries on every side came gusts of fetid air. The streets and alleys reeked with the effluvia of a slave-ship's between-decks. As I passed I had glimpses within of pale babies gasping out their lives amid sultry stenches, of hopeless-faced women deformed by hardship ... (p. 213)

Actually this was not just true of the South Cove, this was the international landscape of victorian urban poverty (Figure 2.1). Andrew Mearns, who worked for the poor of south London and was an influential reformer, described a similar scene in 1883 off Long Lane in Bermondsey, where 650 families lived in 123 houses, their rooms linked by windowless passages, with filth-infested courtyards and four toilets for 36 families (pp. 72-4). Most late nineteenth-century cities had equivalent districts.

Socialism and new technologies

Clearly Bellamy, Morris, and all their contemporaries who dreamed of and acted for reform, had good grounds for doing so. Utopian novels must have given substance to the hopes of

Figure 2.1: International landscapes of urban poverty and street congestion at the end of the nineteenth century: a slum in Stepney in London, and the intersection of Dearborn and Randolph Streets in Chicago. It was on the streets that urban problems were most visible, and many subsequent changes in the appearance of cities began specifically in a reaction against scenes such as these

Sources: Whitehouse, 1980, p. 163, original in the collection of the GLC; Mayer and Wade, 1969, p. 215, original in the collection of the Chicago Historical Society (Number ICHi-04192)

reformers as they took on the daunting tasks of repairing cities and restructuring institutions, not least because these novels were based in the recently formulated philosophies of Karl Marx and of socialism. Edward Bellamy apparently thought of himself as introducing socialism to America, and William Morris was a leading spokesman for socialism in England. 'What I mean by socialism', Morris wrote in 1894 in a particularly fine passage,

> is a condition of society in which there should be neither rich nor poor, neither master nor master's man, neither idle nor overworked, neither brain-sick brainworkers nor heart-sick handworkers; in a word, in which all men would be living in equality of conditions, and would manage their affairs unwastefully, and with full consciousness that harm to one would mean harm to all.

If such socialist reforms seemed likely to be achieved it was partly because contemporary technical advances also seemed to point towards an entirely different society in the future. The 1880s and 1890s saw the development of public sanitation and filtered water supply systems, food preservation by canning, structural steel, asphalt paving for roads, greatly improved elevators, electric street cars, revolving doors, central heating systems, and methods for the large-scale production of plate glass for store windows. The first telephone had been demonstrated in 1876 at the Philadelphia Centennial Exposition and by 1900 there were almost a million of them in the United States, all connected by wires strung between poles, now mostly gone but for over half a century one of the foremost features of the urban scene (Figure 2.2). Such changes were revolutionising everyday life, especially for the middle and upper classes. Out of season fruits and vegetables were being shipped in refrigerated box-cars from regions with different climates, meats kept in cans, products displayed in great store windows, streets made free of at least some of the filth of horses, epidemics reduced and distances collapsed by railways and telephones. Indeed by the end of the century most of the technologies necessary for the tall office buildings, department stores and neat streetscapes of the modern city centre had been invented and were being incorporated into the urban fabric.

Automobiles too were developed in this period, but it was not then clear that they were going to have a profound impact on the

Figure 2.2: Technological innovations of the late nineteenth century suggested an entirely different future world. The beginning of the elimination of distance — a Bell Telephone advertisement in 1895

Source: Atwan *et al.*, 1979, p. 127

form and appearance of cities. At first they were only toys for inventors and the very rich, they could travel little faster than a horse and carriage and had the added inconvenience of frequent breakdowns. In the *Scientific American* for July 1899 and advertisement for a Winton gasoline automobile declared simply 'Dispense with a Horse', but it was over a decade before it came to be widely accepted that cars were going to serve much more than a recreational purpose. Initially they had to fit in alongside the horse-drawn carriages and electric streetcars on dirt roads and cobbled streets that had no traffic lights, pedestrian crossings, roundabouts, direction signs, or parking facilities, and few traffic regulations of any kind.

It was, however, commercial electricity which did most to inspire visions of radically different future cities. Electricity had a magical quality. It was so new and so clean compared with gas and coal, its processes so invisible, that its possibilities seemed to be almost unlimited. Early experiments with electric arc lamps, including a flood-lit soccer game watched by 30,000 spectators in Sheffield in 1878, and streetlamps installed in London between Westminster and Waterloo and in Cleveland in 1879, were dramatic but had demonstrated that these lamps gave off an intense glare and therefore had limited uses. However, the invention in the late 1870s, more of less simultaneously by Joseph Swann in England and Thomas Edison in America, of an incandescent light bulb suitable for both indoor and outdoor use was immediately recognised as being of great significance (Figure 2.3). Regular train excursions were organised from New York City to Edison's laboratory in New Jersey, where thousands of spectators spent a happy evening staring at a display of light bulbs. Concerns about the possible damaging effects on the eyes of this new

Figure 2.3: The marvel of electricity in the 1880s. Looking at the electric light at the Mansion House in 1881

Source: *Illustrated London News*, Vol. 1, 1881, p. 349

form of light were laid to rest in 1882 by a commission of specially appointed physicians, and thereafter it was clear that incandescent bulbs gave electricity commercial and municipal value on a large scale. Indeed one of the terms of reference Edison had given himself for his electrical inventions was that they should be cheaper than gaslight. He quickly followed up his initial success by designing a commercial generating station and by the end of the 1880s such stations had been installed in several American cities. Equivalent stations were also operating by then in some English cities. In 1889 the first electrical streetcars were introduced, by 1900 they had already replaced most of the horse-drawn streetcars. In the 1890s the clean energy of Niagara Falls was harnessed for electrical production, and the engineers who controlled the generators by means of mysterious dials and switches were housed far from public view in neo-classical buildings — the high priests in the inner sanctums of temples to energy. 'It is a new century,' the historian Henry Adams (p. 301) wrote in a letter in 1900, 'and electricity is its god'.

In *Equality* (1897), an uninspired sequel to *Looking backward*, Bellamy wrote that 'Even before 1887 ... the possibilities of electricity were beginning to loom up so that prophetic people began to talk of the day of the horse as almost over', and he envisaged electrical flying machines, 'electroscopes' (televisions), and electrical cars travelling swiftly on roads half the width of the roads of 1890 (impeccable logic — if the same number of vehicles travel at twice the speed only half as much space is needed for them). In this case Bellamy's imagination ran only a short way ahead of reality. By the end of the century daily life was being rapidly electrified. An 'Electrical Club' for the devotees of electricity was established in New York City, replete with all the latest gadgetry of electrically operated doors, stoves, elevators, dumb-waiters and lights. Domestic electrical machines such as toasters and vacuum cleaners and washing machines were being patented, and the flickering gas lights of city streets were being replaced by the steady glare of electrical lamps and the flashing seductions of advertisements. In 1893 on Broadway in the vicinity of Greenwich Village the first electrical advertising sign had been turned on: 'Buy Homes on Long Island', it exhorted, and then that message in bulbs was extinguished and a new one appeared — 'Swept by Ocean Breezes'. The electrical landscape had been invented. There would be no going back.

Edward Bellamy's Boston in 2000

In *Looking backward* an insomniac is put into a hypnotically induced sleep in a sound-proof underground chamber in his house in Boston in the year 1887. The house burns down above him, he is presumed dead, but stays in his trance until 2000 when his chamber is uncovered and opened. Through a series of conversations with the members of the family who awakened him, and some brief expeditions around the new Boston, he learns of the society of the future.

Bellamy was explicit that his account of this society was not to be taken as some fanciful entertainment; for him it was 'a forecast of the next stage of the industrial and social development of humanity' (p. 220). In this stage all capital will have been consolidated into a single syndicate representing all the people so that the nation can operate with maximum efficiency for the sake of common interest. While this requires the submission of individual desires to the whole society this is given willingly, for the nation does everything benevolently and reasonably. There is full employment, everyone being appointed to a position in the 'industrial army' on the basis of aptitude, though they must first spend three years doing manual tasks. Incomes are guaranteed, equal, and ample for all reasonable needs.

This well-ordered, efficient society is housed in an equally well-ordered landscape. Boston in the year 2000 consists of

> miles of broad streets, shaded by trees and lined with fine buildings, for the most part not in continuous blocks but set in larger or smaller enclosures ... every quarter contained large open squares filled with trees ... public buildings of a colossal size and architectural grandeur unparalleled in [1887] raised their stately piles on every side (p. 43).

This is an electrical environment, clean and sparkling, its streets filled with trees and fountains, its buildings adorned with statues representing Plenty, Efficiency and Industry. There are no chimneys and therefore no smoke; there are also no prisons, no trains, no banks because money has been replaced by a universal credit card system; there are apparently no horses because people walk everywhere; there are no stores and therefore no window displays and no advertising. Many of the colossal public buildings are communal warehouses, one for each neighbourhood so that no house is more than ten minutes walk away, and in these are displayed samples of every available product; a purchase is made from the sample by means of a credit card and the item is delivered to one's house either by vacuum tube or to the door. Other public buildings have meeting rooms, recreational facilities and communal kitchens and restaurants where most people eat their evening meal. People walk to these from their 'simple detached dwellings, each standing in its own enclosure'. If it rains or snows the sidewalk is protected by a continuous awning which unrolls automatically.

Bellamy's Boston is a low-density anticipation of a sort of Garden City Beautiful, and it may well have affected the conception of both of these. In *Equality*, Bellamy's second book, the hero takes a trip across America in an electrically powered flying machine and can see that cities have such ample parks and gardens that even Manhattan looks like a large village. This dramatic change from the congested cities of the nineteenth century has occurred because it has been recognised that cities were devouring the countryside and were acting 'as whirlpools which drew to themselves all that was richest and best, and also everything that was vilest', such as crime and poverty (Bellamy, 1897, p. 290ff). Accordingly, the surplus population of cities has moved back to the country because there could be found all the economic benefits of good employment and good dwelling. The villages offered the same communal facilities as the cities, including vacuum tube delivery systems, and in fact the benefits of them were so great that the cities themselves were remodelled along the low-density lines of the countryside.

William Morris's future English landscape

Morris was deeply upset by the excessively centralised and mechanical vision of a socialist future that Bellamy offered, and wrote *News from nowhere* to present his own idea of a decentralised socialist society set in the landscape of southern England in the twenty-first century. This England would be a handicraft, communal, co-operative sort of place, in which both the institutional and physical evidence of industry has been completely dismantled. Machines are used only for the most tedious work, leaving all pleasurable tasks to be done by hand. There are no centralised authorities, no poverty, no exploitation. A bucolic landscape has been created, and this is described in the course of a journey through London and then by boat to the upper Thames valley.

London has become a cluster of villages separated by pleasantly tamed woods. A few old buildings remain as monuments to the previous age, the British Museum, for example, and the Houses of Parliament which 'are used as a sort of subsidiary market and a storage place for manure'. Around Trafalgar Square the neo-classical buildings, which Morris disliked, have been replaced by 'elegantly built, much ornamented houses', each standing in 'a garden carefully cultivated, and running over with flowers'. The only substantial public buildings are the ornate and skilfully decorated market halls. There are no signs of commercial activity because, as in Bellamy's Boston, money is no longer used; in a short street of 'handsomely built houses' with 'an elegant arcade to protect foot-passengers', that is on the site of Piccadilly, there are stores displaying wares but there is no charge for any of these, one simply takes what one needs. At a larger scale the industrial cities have been cleared, Manchester has ceased to exist and has reverted to fields and farms, other cities have been reduced to small towns or rebuilt to rid them of the ugly traces of the nineteenth century. Villages, however, have grown, for in the post-revolutionary era people flocked to them and soon found satisfactory occupations as craftsmen.

> The town invaded the country, but the invaders ... yielded to the influence of their surroundings and became country people; and in their turn as they became more numerous than the townsmen, influenced them also; so that the difference between town and country became less and less (p. 60).

England, in Morris's vision, would have become by the end of the twentieth century 'a garden where nothing is wasted and nothing is spoilt, with the necessary dwellings, sheds and workshops scattered up and down the country, all trim and neat and pretty' (p. 61). The buildings would be exquisitely crafted, decorated with statues and carvings, their style embracing the best qualities of the gothic of northern Europe with those of the 'Saracenic and Byzantine', though none of it directly copied. The overall landscape is readily imagined for it is clear from Morris's description that it would be the familiar landscape of old rural England:

> A delicate spire of an ancient building rose up out of the trees in the middle distance, with a few grey houses clustered about it; while nearer to us ... was a quite modern stone house — a wide quadrangle of one storey, the buildings that made it being quite low ... it had a sort of natural elegance, like that of the trees themselves (p. 161).

News from nowhere probably depicts Morris's most idealistic view of the future, a future based in fact on a rediscovery of the best features of the Middle Ages, which Morris greatly admired, combined with some technological advances, all accommodated to a gentle, decentralised socialism. Elsewhere in his writings he speculated about less dramatic changes that might occur. He imagined, for instance, that in London there might be tall blocks of apartments — he called them 'vertical streets' — with common laundries and kitchens and public meeting rooms (cited in May Morris, 1966, pp. 127-8). These would, he realised, have to be widely spaced so that each could have its share of pure air and sunlight, and garden space and good playgrounds (all principles restated by the Bauhaus and Le Corbusier in the 1920s). Though Morris hoped that reason and good taste would prevail in the design of these, he also realised, with remarkable prescience, that such buildings could well become barren, prison-like settings for their inhabitants.

Utopian realities

The landscapes of the 1980s have not turned out as either Bellamy or Morris hoped. In their utopias neither money nor large cities were needed. In reality capitalism and com-mercialised materialism have flourished, cities have expanded upwards and outwards, the socialism that has been most widely adopted is of the dull centralised type, and the technologies that have changed the look of the world have encouraged massiveness rather than personal responsibility and decorative art. Neverthe-less, between them these two authors did anticipate many of the features of the modern welfare state, including socialised medicine, nationalised industries, completely planned towns,

Figure 2.4: London and Boston today, not as Morris and Bellamy imagined they might be. London's skyline looking north-west from The Monument, and Route 1 in the northern suburbs of Boston

and the widespread use of electricity. More specifically and tangibly, the architects of the first garden cities at the turn of the century, and of the Bauhaus in the early 1920s, paid tribute to the ideas of Bellamy and of Morris, and in the former case seem to have incorporated some of them into their designs.

Nevertheless, the fact is that the appearance of modern cities owes almost nothing to Morris and Bellamy. If, through some local accident of physics, they were to be time-transported into London and Boston now they would be profoundly disappointed by almost everything (Figure 2.4). The unornamented sky-scrapers of banks and insurance companies, the new roads scarcely less congested than the old ones, the same ugly buildings around Trafalgar Square, sex shops, numerous clusters of bare and prison-like 'vertical streets', the great bland offices of centralised government in downtown Boston, and the Houses of Parliament still not used for the storage of manure.

Exact anticipations would in fact have been little more than coincidence even if they had been right in almost every detail, and it is not very important whether Morris and Bellamy got them right or wrong. What is significant about these two utopian books is, first, that they remind us that there have been few inevitabilities in the making of the modern city; if other attitudes had prevailed cities now could have very different landscapes. And secondly, they made popular a way of thinking about future societies and landscapes as something other than an extension of an existing state of affairs. This idea is now so commonplace that we take it for granted, yet in the 1890s, after centuries of looking to the past for inspiration, it was indeed a radical insight. It was taken up quickly and enthusiastically by planners and architects, at least some of whom managed to manoeuvre themselves into positions from which they could turn their own utopian conceptions into built realities. Garden cities, the Deutscher Werkbund, the Futurists, the Bauhaus, La Ville Radieuse of Le Corbusier, Frank Lloyd Wright's Broadacre City, neighbourhood units, even mundane municipal plans, have all been informed by the conviction that the present problems of cities can best be transcended by looking to the future.

3

Old Styles and New Forms in Architecture: 1880-1930

In 1884 construction was completed on the Home Insurance Building in Chicago. From the outside it was not unlike many other commercial buildings of the period, a big block of a structure built of stone in a mixture of styles. Its height at ten storeys was unusual, but not exceptional. What was remarkable about it was its structure. The walls were supported by a metal frame. For the lower six storeys this was of wrought iron, a construction material that had been used in various ways for several decades; but the top four storeys were held up by a skeleton frame of steel, the first known use of structural steel. With its great strength relative to weight the steel skeleton frame was to change the nature of architecture: it made skyscrapers possible, it contributed to the invention of an original architectural style, and dramatically altered the character of the urban landscape.

The Home Insurance Building presented an architectural paradox that was to last well into the twentieth century (Figure 3.1). Quite simply, the new methods of construction, which made all sorts of new forms possible, were disguised behind a traditional façade. Most architectural histories treat the development of the modernist, angular, undecorated style, that eventually came to dominate office and institutional building in the 1960s and 1970s, as though it has been a smooth and rational progression from structural steel to reflecting glass cubes. Such histories are at best partial accounts of what happened. For much of the last hundred years it was by no means certain that undecorated modernism would triumph as it did, and urban landscapes bear substantial evidence of the enduring popularity of ornamental styles. These styles declined slowly, getting progressively simpler and more truncated as modernism became more popular, until they finally fell away in the 1950s.

Figure 3.1: The Home Insurance Building. Designed by William le Baron Jenney, and built in Chicago in 1881, this was the first steel skeleton skyscraper, a fact disguised by the conventional decorative façade

Victorian architecture

For all the utopian speculation, technological innovation and economic growth in the nineteenth century, or perhaps in reaction against them, victorian builders clung to well-tried appearances for their buildings. Steam engines and the machines of mass production were invariably housed in structures that looked like botched versions of greek temples or medieval cathedrals. By 1880 almost every sub-style of architecture had been revived, modified and combined with all the others. In the early years of the twentieth century classical decoration had a resurgence; and in the 1970s and 1980s there has been a further revival of elements of older styles. The result is that it is impossible to understand the architecture and landscape of the twentieth century without at least a scant knowledge of architectural history.

Western architecture up to 1800 took four major forms — vernacular, classical, gothic and renaissance. The regionally varied vernacular styles were made by craftsmen using local materials and traditions; these included, for example, the stone buildings of the hilltop towns of Provence, and the half-timbered structures of the Wales-England border country. Though they have often been a source of inspiration for suburban house designs, until recently vernacular buildings were ignored by architectural historians, who contemptuously considered them to be crude and quite inferior to real 'Architecture'.

Architecture, in this high or snobbish sense, began with classical styles, those of the greek temple and the roman stadium. These were distinguished by their low pitched roofs, symmetrical facades, mathematical proportions and carefully prescribed geometries of right angles and half circles expressed in precisely defined 'orders', each of which carried with it specific types of columns and statuary. In the Middle Ages these were displaced by the gothic styles of churches with their pointed arches, spires, towers and steeply pitched roofs reaching heavenward. Such buildings were a relatively free expression of the skills of master builders and the craftsmen who covered them in ornate carvings, and of the spiritual values which pervaded medieval society. Renaissance architecture, in contrast, was primarily a revival of classical styles, albeit adapting them to new functions such as churches and palaces, and modifying them by the addition of new forms of decoration and windows. Renaissance styles dominated public buildings from about 1500 up to the beginning of the nineteenth century, and in simplified forms they also became very popular for the houses of georgian England and colonial America.

As the eighteenth century turned into the nineteenth there was a widespread reaction against all forms of classicism, and a plea for more romantic, spiritual and emotional types of art, poetry and architecture. The architectural style which suggested itself was gothic, and as a result of a self-conscious effort by individuals such as A.W.N. Pugin, the architect of the English Houses of

Parliament, and the art critic John Ruskin, gothic styles were revived. Ruskin's books were widely read in both England and North America; he fiercely criticised the machine-like work of machine industry, deplored the ugly environments in which it took place, and made an impassioned argument for the free, inspired craftsmanship that he identified in the buildings of the Middle Ages. His complex social and aesthetic arguments were soon simplified and popularised into the idea that a building with gothic trimmings was beautiful and romantic regardless of what went on within, perhaps because the owners hoped that a spiritual style would compensate for their entirely mercenary purposes. Neo-gothic façades were soon adopted and adapted for every possible purpose — railway stations, offices, houses, law courts, town halls, schools, and museums (Figure 3.2).

Figure 3.2: Nineteenth-century architectural revivals. a) The high gothic of the Law Courts on Fleet Street in London; b) the plain gothic of row houses in Toronto; c) a Romanesque (or Richardsonian) house in New London, Connecticut, is characterised by its bulkiness and rounded arches; d) a Queen Anne house, with tile hung walls, a round tower, a variety of window shapes, brick and carved stone, probably built in the late 1890s, Toronto (and which has recently been chopped vertically in half and moved a hundred feet to accommodate new construction)

The gothic revival opened the floodgates; if one old architectural style could be revived so, it seemed, could all the others. First of all, classical styles were resurrected, especially for banks and government offices, though the strict rules of classical architecture were often broken and symmetry was ignored in order to fit buildings to constricted sites. Then, in the final quarter of the century, revivals and modifications came faster and from more varied sources. The Second Empire or Mansard style, originally used in Paris as a means of complying with strict building codes about height by carrying usable space up an extra storey while giving the appearance of a conventional roof, became internationally fashionable in the 1870s; Dutch gables were widely used for a few years; heavy-set and turreted italianate buildings were common in the 1880s; romanesque revivals with broad round arches over windows and doors were popular in the 1890s. The Queen Anne styles (the origin of the name is obscure) in vogue at the end of the century were a clear expression of the confusions that resulted from all these revivals, for they combined elements of most of the other styles, with turrets, romanesque doorways, tile-hung walls, odd classical pillars, mixed materials and windows in a variety of shapes and sizes.

Looking around him in 1901 Frank Lloyd Wright was not impressed by the mixed up urban landscape that had been created by all these revivals. He could see little to praise and much to criticise, especially the surfeit of trivial differences that resulted in what he characterised with astute precision as 'monotony-in-variety'.

The decline of the last classical revival

In the last years of the nineteenth century gothic architecture faded in popularity. Perhaps the dark, spiritual forms were considered to be inappropriate for an electrical age; perhaps they had just been done to death and people wanted no more of them. Whatever the case, only a handful of neo-gothic buildings were constructed after 1900, one of the last being the Palace of Peace at the Hague, ironically opened in 1914. What replaced the gothic revivals, and persisted long enough to determine the visual character of, for instance, much of present-day central London, were modified classical styles. At first these were bedecked with cherubs, columns and all the details of the various orders, and they were used on every sort of building from government offices and department stores to generating stations and houses; then they were gradually simplified and the decorative features were omitted; by 1950 the qualities of classical architecture had been reduced to little more than a suggestion of columns and proportion on a few new buildings. Nevertheless it is probably accurate to say that in the first half of the twentieth century classical revivals were the most fashionable of all architectural styles. Their impact on modern urban landscapes is considerable.

The twentieth-century classical revival was both a European and a North American phenomenon, but its single most powerful impetus may have come from the Ecole des Beaux Arts in Paris, the values of which held sway in architectural design teaching almost everywhere around the turn of the century. The Beaux Arts philosophy was to challenge careless eclecticism by ensuring that new buildings were precisely accurate in copying past styles, and especially to encourage the copying of renaissance models. The graduates of the Ecole des Beaux Arts carried this philosophy with them wherever they went, and that included North America as well as Europe.

In America classical revivalism was more specifically stimulated by the Columbian World's Fair held in Chicago in 1893. This fair took the form of a huge exposition, officially to celebrate the four hundredth anniversary of Columbus's discovery of America, but in reality to display the technical and scientific achievements of the age (though it also had an extensive midway, a precursor of Disneyworld, with a replica of a Bavarian castle, a copy of Battle Rock Mountain in Colorado, and a Polynesian village complete with villagers). Previous exhibitions had set high standards — Paris with the Eiffel Tower in 1889, Philadelphia with the first telephone in 1876, and of course Crystal Palace and the Great Exhibition of 1851. The man in charge of the works at Chicago was Daniel Burnham, a local architect who had designed a number of banks, offices and schools. He was convinced that the outstanding feature of this exposition should be its architecture, and that this should be classical, with every building standing alone to display the grandeur, elegance and other qualities of classical civilisation that were supposed to be being reawakened in the United States of America. And thus it was built as a glorification of classicism. There was a great peristyle, building after building with domes and balustrades and porticoes, there were statues of Justice and Plenty and Industry, classical orders veritably abounded (Figure 3.3). The exhibition was proclaimed as the City of Palaces, the City of Light (it was the first exhibition lighted by electricity), the White City. Popular enthusiasm for it in America knew few bounds; there were 21 million visitors; journalists extolled it and saw in it a microcosm of a socialistic planned society, and one even claimed that it 'revealed to the people possibilities of social beauty, utility and harmony of which they had not been able even to dream' (cited in Hines, 1974, p. 120). Here, it seemed, was the future world made present.

Perhaps it was. The Chicago Fair undoubtedly left an enduring mark on American landscapes. It became inconceivable in the following decades for state capitols or universities to be built in anything other than some adaptation of classical architecture with its associations of democracy and reason. Classical was chosen for almost all other government and business build-

Figure 3.3: A corner of the World's Columbian Exposition, Chicago, 1893, the event which stimulated the popularity of classical revival architecture for the next 40 years. The statue of the bull, on the left, represented 'Plenty', and the boy with the horse, in the centre, represented 'Industry'; the significance of the moose is unknown

Source: *Picturesque World's Fair*, p. 93

31

ings too, including the Wall Street Stock Exchange, the New York Public Library, the tops of skyscrapers, most post offices, and even the little temples to electricity which housed the equipment for the new generating stations at Niagara Falls.

Popular though it was, not everyone greeted the neo-classical style of the Columbian Exposition with unbridled enthusiasm. Even Edward Bellamy, with whose socialist vision of the future city the exposition would seem to correspond quite closely, found it too ornate and superficial. French critics dismissed it scathingly as nothing better than a set of student exercises from the Ecole des Beaux Arts. Louis Sullivan, who had designed one of the few buildings on the grounds which was not in a classical style, subsequently commented that 'the damage done by the World's Fair will last for half a century from its date, if not longer'. Montgomery Schuyler, probably the leading architectural critic of the 1890s, agreed. In a carefully argued criticism of the fair published in 1894 he maintained that it was based on unity, magnitude and illusion, chiefly illusion. The unity was achieved not by the common architectural style but by the excellently conceived and executed landscape plan (the work of Frederick Law Olmsted, the designer of Central Park); the magnitude was because it had displayed the biggest of whatever it could; the illusion was in the display of the electrical lights and electrically-powered fountains, but especially in the architecture taken from another age, conceived for a time when life was different, an architecture that had nothing whatsoever to do with the machines it housed. In the context of late nineteenth-century industrial society classical revival styles were not merely inappropriate, they were fraudulent. And for forward looking architects they weighed like a stone around their necks. Frank Lloyd Wright later declared that the World's Fair was 'a mortgage upon posterity that posterity must repudiate not only as usurious but as forged' (cited in Mumford, 1952, p. 171).

The criticisms were to little avail. In Europe as in America classical revival styles were fashionable, and anyway there seemed to be no alternative to them. They were used for many of the buildings constructed in Paris just after the turn of the century, including, for example, Sacré Coeur which was completed in 1914. In London new government buildings in Westminster, the Kingsway-Aldwych redevelopment, the modifications to Regent Street, the great department stores on Oxford Street, County Hall, were all built in various renaissance revival styles before

World War I. Then in 1914 Geoffrey Scott published an influential book, *The architecture of humanism,* in which he made a convincing argument for the merits of renaissance over gothic architecture, thus making explicit in words what was already widely apparent in urban landscapes, namely that an adapted classical architecture was the one that would be most appropriate for the rational and humanistic society of the twentieth century.

But the fact was that by then such attachment to the architecture of classicism was becoming contrived, especially for large public buildings. Even in the 1890s some of these had been draped over steel frames and by 1910 this had become the common practice both in Europe and America. Classical styles had, in effect, already become no more than superficial ornament, columns were no longer needed to hold buildings up, they just looked nice and conveyed the right suggestions. The carved decorations and statuary were expensive, and possibly as a cost-saving measure they became less and less ornate. After World War I the simplified, bas-relief decorative styles of Art Deco became fashionable, and when classical style features were used they were so muted that by 1930 they had been reduced to little more than slim decorative pilasters, a balustrade with a few urns and some etched panels. There was a brief, infamous moment of recuperation when the National Socialists in Germany and Italy built great neo-classical monuments to themselves at Nuremberg, Rome and elsewhere. Otherwise this process of simplification continued almost as though it was a preordained devolution. In buildings of the early 1950s it is occasionally possible to see remote suggestions of classical forms in the arrangement of doorways and windows, by the end of the decade even these had fallen away. After half a millennium as a fashionable architecture for public buildings classical styles had finally faded from view, slowly, like the grin on the Cheshire cat (Figure 3.4). Hardly anyone noticed their passing.

Building big and building tall: Crystal Palace and the Eiffel Tower

As classical styles declined new forms of building arose to take their place. Easily the most spectacular of these was the sky-scraper. This was the most visible product of the new technologies of the late nineteenth century, for it would not have been

Figure 3.4: Evidence of the slow decline of classical styles in the twentieth century. a) Wall Street Stock Exchange, 1903; b) County Hall, London, 1920s; c) US Post Office Building in Oak Park, Chicago, 1933 (opposite Frank Lloyd Wright's Unity Temple); d) west wing of the National Gallery, Washington DC, 1941; e) Ontario College of Pharmacy, 1940

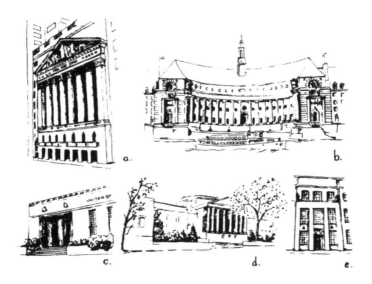

possible without structural steel, nor without electricity to power the elevators and to light the offices; nor indeed would sky-scrapers have been profitable or made business sense had it not been for the rapid growth of corporations and the development of typewriters and telephones which were changing the function and operation of offices. Skyscrapers are in fact so tied up with technology and with business that they have become a demonstration of the meaning of progress and the foremost symbol of capitalism.

Prior to the nineteenth century large buildings invariably expressed religious or political authority, they were temples, churches and palaces. Skyscraper building, however, seems not to be just a simple extension of this desire to demonstrate authority, but appears to have emerged largely out of the Victorian engineer's mundane inclination to build things big and tall for little more purpose than to show off technical prowess. This was the case, for example, with Crystal Palace. Erected in Hyde Park to house the Great Exhibition of 1851, it was a symbolic 1,851 feet long, tall enough to enclose mature elm trees, made from three modular prefabricated sections, and took just six months to construct from the first notepad doodle to opening the doors to receive the first exhibits. It was an engineering demonstration of the structural strength and speed of construction that could be achieved through precision and mass production. The Eiffel Tower, built for the Paris International Fair in 1889, was justified as a tower of pleasure and as a means of sending military communications by helioscopes (flashing light off mirrors), but, as the name indicates, was in fact little more than a display of the design skills of the engineer, Gustav Eiffel. Nevertheless, to late-Victorians this wrought iron structure rising over 1,000 feet seemed to symbolise man's superiority over nature through the application of science and engineering. Thomas Edison, no doubt bathed in an afterglow of light bulbs, looked upon the Eiffel Tower and thanked God for 'so great a structure'. Not everyone was so overwhelmed. William Morris declared that whenever he visited Paris he tried to stay as close to the base of the tower as possible, for then he could scarcely see it. His was undoubtedly a minority opinion.

Early skyscrapers in Chicago

The Eiffel Tower was the last of the big for the sake of bigness structures, because by the 1880s it was already possible to construct buildings both tall and useful. Throughout the century commercial and industrial buildings had been getting larger. With the invention of the first safe elevator by Elisha Otis in the 1850s and its subsequent adoption in office buildings and hotels, the effective height limit of about five storeys, imposed by how many flights of stairs a customer was willing to climb, was broken. The Tribune Building completed in 1873 in New York City had nine storeys plus a decorative tower, and soon there were many elevator buildings which seemed to tower over their five-storey predecessors. Related improvements in construction techniques made possible the erection of the 16-storey, all-masonry Monadnock Building in Chicago in the late 1880s. But such an all-masonry structure has to have enormously thick walls at the ground level to support the weight of all the bricks and mortar above; it is not very economical or efficient. In any case by 1889 all-masonry construction was obsolete. The skeleton steel frame with masonry cladding had been invented in the Home Insurance Building built in Chicago earlier in the decade, and was rapidly asserting itself as the best way to build large structures.

In the 1880s and 1890s Chicago was the centre of a great conflation of architectural developments. Indeed, one of the reasons why the 1893 Columbian exposition was criticised for being retrogressive was that there were so many signs of local progress in a quite different direction. It was in Chicago in the 1880s that the modern office skyscraper was invented. It is not wholly clear why this happened there rather than elsewhere, but possible reasons include the great fire there in 1871 which had opened the way for a building boom, and rapid population growth — between 1870 and 1910 it increased from 325,000 to 2,100,000. This growth pushed up land values — they increased by a factor of seven in the 1880s alone — and this may have made it more profitable to build vertically rather than horizontally. At the same time there was a strong demand for pretentious office space from big new corporations like the railroad and meat packing companies. Furthermore, skyscrapers turned out to be very profitable investments. It is unlikely, however, that Chicago was unique in these respects, so there seems to be no clear reason why skyscrapers should have been built there before anywhere else. One can only speculate that Chicago's developers had few of the constraining pretensions of the capitalists of other major business centres; they saw architectural embellishment as a waste of money, they wanted value for their dollars, and they were willing to experiment with the new steel-frame skyscrapers to demonstrate their progressiveness.

This experiment firmly established what Lewis Mumford has called the architecture of 'the steel cage and the curtain wall' — curtain wall because the walls no longer have to be load bearing, they are little more than curtains to keep out the inconsistencies of weather. From this much of modern architecture has developed. The advantages of the curtain wall skyscraper were immediately apparent to astute Chicago businessmen. It had the great merit of increasing room size, especially on the ground floor where the rents were highest and conventional masonry walls would have been thickest (Fryer, 1891). Building up was clearly the way of the future, and in the first volume of the *Architectural Record*, published in 1891, an enthusiastic correspondent felt free to declare that 'High buildings are demanded and today there is simply no limit to the height that a building can be safely erected'.

Skyscraper styles to 1930

It remained to work out styles which were suitable for these new architectural forms. This proved to be a good deal more difficult than their structural development.

The first architect to conceive a distinctive style for skyscraper architecture was Louis Sullivan, working in conjunction with his partner Dankmar Adler. Based in Chicago he designed a number of buildings both there and in other Midwestern cities which honestly and clearly expressed the steel frame beneath their stone and brick cladding, and which architectural historians consider almost unanimously to be high architectural achievements. The Guaranty (renamed Prudential in 1895, but generally known by its original name) Building in Buffalo is a fine example of Sullivan's work (Figure 3.5). It has three clearly defined components: a base for shops, a shaft for offices, and a capital to screen elevator machinery. Strong vertical lines reveal rather than disguise the steel frame structure. In detail its surfaces are intricately decorated with a kind of terracotta filigree. Sullivan saw his buildings as organic expressions of nature manifesting itself through structure and decoration; the concave cornice on the Guaranty Building was for him the point where the 'life force' on the surface expands in swirls and so expresses the mechanical system of the building which 'completes itself and makes its grand turn, ascending and descending'.

Around the back this organic philosophy is not quite so clear. There the building is entirely unornamented except for a fading Prudential sign painted on the exposed brickwork, and there is a great cleft to allow light into the offices. There are 23 offices to each floor, and for all the architectural accomplishments this was, as Sullivan well realised, just an office building: 'an indefinite number of storeys of offices piled tier upon tier', he wrote, 'one tier just like another tier, one office just like all the other offices — an office being similar to a cell in a honey-comb, merely a compartment, nothing more'.

Sullivan's designs exercised little influence over his contemporaries. J. Schopfer, writing in the *Architectural Record* in 1902 (p. 271), declared somewhat wistfully that 'Even in the United

Figure 3.5: Three views of what is known to architectural historians as the Guaranty Building, Buffalo, by Adler and Sullivan, 1895, even though it has several signs declaring it to be the Prudential Building. An early skyscraper with a form which clearly reflects the steel skeleton within

States architects in the majority of cases have not been able to bring themselves to proclaim to the man in the street that the sky-scraper has a framework of steel; they have given these twenty storey edifices the aspect of houses, which is an absurdity.' Absurdity it might have been, but there was no denying its popularity. From about 1890 to 1916, and in some cities up until 1930, the common aim was to build skyscrapers in the revival fashions of the time. This was done by the simple practice of taking a standard Beaux Arts or equivalent building, splitting off all the roof details at about the second floor and hoisting them up 20 or 30 storeys on a rather plain sort of column. The decorations at ground level could be seen, those in the sky could not be seen except by some neck-craning effort, and the column presumably was not meant to be seen at all. The results of all this can be found in any city where tall buildings were constructed in the first half of the century; they are accurately described as stretched temples, elevated palaces and elongated cathedrals (Figure 3.6).

These ornamented skyscrapers were also characterised by increasing height — 30, 40, even 50 storeys. The Woolworth Building in Lower Manhattan was completed in 1913 as a great gothic elongation, 52 storeys and 792 feet tall. It was for many years the tallest useful building in the world. In 1913 it was also voted by some committee or another the most beautiful building in the world. Not everyone would have agreed. Henry James, returning to America after many years in Europe, was deeply dismayed by the skyscrapers of Manhattan. In *The American scene* (p. 76) he wrote of 'the multitudinous skyscrapers standing up to the view ... like extravagant pins in a cushion already overplanted, and stuck in as in the dark, anywhere and anyhow ...' They were much too vulgar and economic for his europeanised taste. On the inside he found 'huge constructed and compressed communities' dedicated to making money and nothing more, their ornate displays a testament to loneliness.

Most critics of skyscrapers had more practical concerns. The skyscrapers were getting too tall. In both Chicago and New York there had been several attempts to restrict skyscraper height by imposing ordinances, but these restrictions had been easily side-stepped by builders (P.B. Wright, 1910, p. 15). The problem was, as one critic expressed it, 'that the architecture renders meaningless all the architectural values upon which the traditional street architecture has been based' (David, 1910, p. 392); by restricting their height it was hoped that some semblance of human pro-

Figure 3.6: A stretched temple skyscraper: the Chicago Tribune Building, 1923. Skyscrapers with these sorts of ornate tops were popular between about 1900 and 1925

portion for streets might be retained. There were also problems of light and air circulation at the bottom of the built canyons which were being created (Figure 3.7). One building constructed in the early years of the century on Lower Broadway has a 90 per cent site coverage and rises 36 storeys sheer from the sidewalk. It was as though the entire city was being turned into a single enormous thick structure intersected by streets which were no more than utilitarian corridors and air shafts.

New York City responded in 1916 to some of these problems by passing a zoning by-law that was to have a profound effect on the appearance of skyscrapers for the next four decades, and therefore on the skylines of almost all North American cities.

Figure 3.7: The origins and logic of the wedding cake style of skyscraper, as understood in 1921. This sort of stepped-back form was required by the New York Zoning By-Law of 1916, and to some degree it was a feature of the design of skyscrapers everywhere up to the 1950s

AMERICAN
RENAISSANCE

THE FRENCH
METHOD

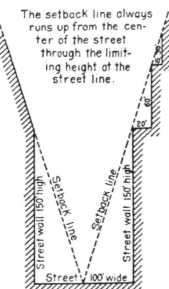

The setback line always runs up from the center of the street through the limiting height at the street line.

Street wall 150 high

Setback line

Setback line

Street wall 150 high

Street 100' wide

THE MODERN
AMERICAN IDEA

Source: Pond, 1921, p. 132

The idea of stepping back the upper storeys of skyscrapers to allow light and air to reach the street had been proposed as early as 1891 by Montgomery Schuyler (1964, p. 387) (and on a somewhat lesser scale it had been incorporated into light and fire regulations passed in Paris and London in the 1890s). In 1916 this step-back principle was entrenched in the New York City by-law, which also restricted skyscrapers to certain specified streets or zones in order to protect residential districts from overwhelming commercial intrusions. The aims of this zoning by-law were to stabilise property values, to relieve congestion in the streets and on transit lines, to provide greater safety in buildings and on the streets, to make the business of the city more efficient, and to make life in the city healthier. In detail the by-law was complicated, and it was changed almost every year for the next 45 years until it was finally repealed. In general and in its architectural effects it was, however, simple — the upper storeys of a building had to be set back from the lot line according to some formula, such as a line drawn from the centre of the street to a cornice line at 90 feet and projected upwards, with no part of the building passing through this line.

Following the implementation of the by-law skyscrapers took on a ziggurat or wedding cake profile. This became widespread as many other North American cities passed similar height and zoning by-laws. The old stretched temples simply could not be fitted into the new requirements, and less ornate styles were soon developed, especially following a design competition for the Chicago Tribune Building in 1922. This competition elicited proposals from both European and American architects, some utterly fanciful, some anticipating the skyscrapers of the 1950s and 1960s. It was the design by Eliel Saarinen which was placed second, and which was never built, that exerted a profound influence on skyscraper styles up to 1950. It had little ornament, a wedding cake outline and powerful vertical lines which seemed to vanish like a spire into the sky. This was as close to a skyscraper style that was consonant both with steel frames and tall buildings as anybody had come since Louis Sullivan; indeed Sullivan, shortly before he died, praised this design as 'a monument to beauty, to faith, to courage, to hope' (1923, p. 157). It was the major design source for the Chrysler Building, the 102-storey Empire State Building, the Rockefeller Center Buildings, and numerous skyscrapers built between 1925 and 1950 in New York and other North American cities (Figure 3.8).

Figure 3.8: The Chrysler Building, New York, 1928-30. A corporate skyscraper in the exaggerated verticality style of the 1920s and 1930s; the gargoyles are huge hood ornaments

The origins of downtown

The concentration of skyscraper offices in city centres both encouraged and complemented the contemporary growth of business corporations. Towards the end of the nineteenth century corporations began to get larger in terms both of capital assets and the scale of their operations. They had money to invest in large buildings; indeed the first use of the neutral word 'Building', as in Home Insurance Building or Chrysler Building, seems to date from this period. With the expansion of operations the head offices of many corporations were severed from their factories and transferred to the centres of the cities, leaving workers in the soot and grime while managers enjoyed all the benefits of big city living. These factoryless businesses had to be closely connected to banks, stock exchanges and related financial services, so these too expanded rapidly, as did related entertainment and retail facilities such as restaurants, theatres, opera houses and big department stores like Harrod's in London and Carson Pirie Scott in Chicago.

George Hill, a contributor to the *Architectural Review* in 1892, gave a simple economic and psychological explanation for what was happening in city centres: tall buildings not only provided more rental space to pay off the initial investment and a superior service for the same outlay as lower structures, but they also answered to the 'desire of all men to be close to the centre, and the desire of men of one calling to be close to one another'. This was the origin of the urban downtown as we might now understand it — a concentrated mixture of offices and facilities for entertainment and shopping housed in large opulent buildings.

A direct consequence of this desire to be at the centre was a demand from the newly affluent businessmen for residences close to their places of work. In downtown areas already largely built up almost the only possibility for this was in apartments. The development of the first apartment buildings closely parallels that of the skyscraper offices. They evolved out of residential hotels which in the second half of the nineteenth century had become seats of luxury meeting all a person's needs with restaurants, laundries, elevators and fully-fitted bathrooms. The first apartments also had their own laundries, and even nurseries for taking care of children; in some of them food was cooked in a common kitchen by professional chefs and served in the privacy of one's own apartment which had minimal kitchen facilities.

The Dakotah Building overlooking Central Park (now fixed in the minds of an entire generation as the building where John Lennon lived and was killed), was one of the earliest of these new apartment buildings. An eight-storey structure, its style has been variously described as Brewery Gothic and Middle European Post Office. It originally had a doorman, sentry box, *porte-cochère*, ample interior courtyard (needed for drying laundry in the days before driers), tennis courts, a restaurant, and in the attic a gymnasium and playground for the children to use in poor weather.

Its elevators were lined with velvet and had cushioned seats. As in offices, elevators had the merit of equalising the value of all the floors, and even rendering the upper floors more prestigious; with elevators higher meant better, not just more stairs to climb.

As more skyscraper offices were constructed so more luxury apartments and hotels were built. They were not as tall as the offices, but in almost every other respect they were similar. They affected the same elongated gothic and renaissance styles, often apparently borrowing them from travel books. Floor plans, like those of offices, were standardised. The overall design approach was simplicity itself. The ground floor was devoted to stores and restaurants; the relatively undecorated shaft was apartments; the ornamented capital was the penthouse suite. This approach persisted into the 1920s. In 1926 Mrs E.F. Hutton (of Woolworth's), as a condition of permitting redevelopment, had her townhouse at 1107 Fifth Avenue demolished and recreated exactly as the top three floors of a 14-storey apartment building on the same site. Such are the whims of the wealthy.

In 1880 churches and the masts of sailing ships still dominated the skylines of New York and other cities. Over the next three decades elevator buildings, then skyscraper offices and apartments created an urban skyline and a downtown landscape that had no precedents (Figure 3.9). Towering 'mountain ranges' of buildings and the new street canyons filled with stock exchanges and department stores were incontrovertible evidence of the technological accomplishments and economic prosperity of a new age. When that prosperity abruptly disappeared in the Great Depression of the 1930s skyscraper construction came to an equally abrupt halt. Though some work was carried out on smaller structures in the 1930s and 1940s, the fate of the Empire State Building served as a stern warning. Its construction spread into the early years of the Depression and it took almost ten years to rent all its floor space; it was nicknamed with sharp accuracy 'The Empty State Building'. Few skyscrapers were to be built, in New York or anywhere else, until the 1950s and 1960s.

European cities had been, from the beginning, much less enthusiastic about building high. Steel frame consruction did not become common in Europe until about 1910, and then for a long time it was masked behind various classical revival façades. Moreover, there were strict height regulations in effect in many cities because of fire and building regulations; for instance the

Figure 3.9: The changing skyline of Manhattan, 1876 to 1931. In 1876 churches were prominent skyline features and Brooklyn Bridge was under construction. In 1913 the skyline was one of stretched temple skyscrapers; the Woolworth Building is on the left. By 1931 the skyline of New York was etched deeply into every architect's psyche and some even dressed as their buildings; the 1931 Beaux Arts Ball in New York, with William van Alen as his Chrysler Building

Sources: T. Adams, 1931, p. 56; *Architectural record*, 1913, p. 99; *Pencil Points.*, 1931, p. 145

Figure 3.10: Brompton Road, London, 1902 and 1931. Between 1880 and 1930 London's skyline changed much less dramatically than those of New York and Chicago, partly because of fire regulations which limited the height of buildings to about the length of an extension ladder. Nevertheless there were substantial changes in the appearance of some streets with the expansion of office buildings and department stores such as Harrod's, which grew from the building with a sign just visible in the top photograph to the great domed edifice which it still occupies

Source: Clunn, p. 334

London Building Act of 1895 specified that structures should not exceed 80 feet to the cornice line with a further two storeys under the roof for a total building height of 100 feet. Combined with deeply established conventions about classical revival styles and with constricted or irregularly shaped building sites, these regulations did much to determine the character of the landscape of central London up to World War II. There were, of course, significant changes in scale and form as eight and ten-storey offices and department stores replaced two or three-storey buildings (Figure 3.10). But building regulations long delayed the construction of tall buildings. There were no skyscrapers in European cities, even on the scale of smaller North American cities like Toronto and Buffalo, until late in the twentieth century. The contemporary developments that took place in European architecture were of another kind, in their way no less dramatic than those of skyscrapers with which they were to be combined after World War II. These were the developments that led to the undecorated geometric styles of modernism which subsequently came to dominate the content of urban landscapes everywhere.

4

The Invention of Modern Town Planning: 1890-1940

In the first 30 years of the twentieth century a number of procedures and ideas for improving urban living conditions were brought together into a coherent system that was called 'town planning'. This term may have been invented by Raymond Unwin, certainly it had not been in use for many years before he popularised it in his book *Town planning in practice*, which was published in 1909. At first town planning was conceived as a way of providing grand solutions to all urban problems, either by radical redevelopment for city beautification or by the construction of entirely new garden cities. It soon became clear that such utopian solutions were not going to be implemented on more than a limited scale, and planners, being practical people, reduced their sights to finding ways of separating incompatible land uses and designing good residential neighbourhoods. It is chiefly these that have left a mark on the patterns and forms of modern cities.

Precedents for modern town planning

From some perspectives, especially its results, town planning looks as though it is an attempt to make cities function as efficiently as factories. The more conventional view is that it began from the opposite motive, as a reaction against the industrialisation which had created such great inequalities in living conditions by exploiting for profit whatever did not have to paid for directly, such as housing, air, water and workers' health. This reaction initially took four rather different forms — municipal by-laws to govern building standards, picturesque town layouts,

49

Haussmann's reorganisation of Paris, and model industrial towns. These were the precursors of modern planning.

Regulations to govern building practices, especially with regard to fire and safety matters, had been in effect for several centuries, and in the second half of the nineteenth century these were greatly extended in order to restrict the unscrupulous practices of builders. At first the new by-laws set minimum standards for window size, doorway heights and so on, but by the end of the nineteenth century they also often specified street widths, backyard size and building height. Though these regulations were intended to improve the design and layout of housing they had the unfortunate consequence of encouraging the construction of monotonous rows of identical houses at 50 to the acre, with no parks or provisions for shops and schools. Large areas of London and Chicago, for example, were developed in this tedious manner of so-called 'by-law planning' (Figure 4.1). In his book on planning Unwin wrote (p. 4) that 'The truth is that we have neglected the amenities of life. We have forgotten that endless rows of brick boxes, looking out upon dreary streets and

Figure 4.1: By-law planning. The map of Fulham in Unwin's book Town planning in practice, *showing the monotonous streets that resulted from the application of uniform building by-laws*

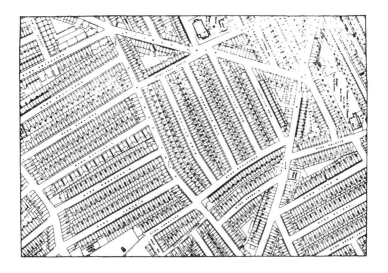

Source: Unwin, 1909, p. 5

squalid backyards, are not really homes for people.' Early twentieth-century planning sought to correct these practices.

The ugliness of industrial cities had in part prompted the gothic revival in architecture, and this duly found its equivalent in planning with the publication in Germany in 1889 of Camillo Sitte's book on town building as a work of art, in which he argued for a picturesque approach to town design, with winding streets and irregular clusters of buildings around town squares. According to this approach towns should expand by a series of concretions, picking up the patterns of existing streets and housing layouts and extending them. In Germany, where municipal controls were stronger than elsewhere, these ideas were quickly applied. They appealed to Unwin, who stressed the role of what he called informal beauty and the need for towns to be 'an outgrowth of the circumstances of the site' (p. 138).

The great Victorian model for urban redevelopment, however, was not Sitte but Baron von Haussmann's reconstruction of Paris in the 1850s and 1860s. Under the authority of Napoleon III Haussmann laid out the avenues, boulevards and major urban parks which give central Paris its distinctive character. At the same time he installed water supply and sewerage systems, and established strict guidelines for building design. All this was achieved by the simple expedient of imposition. The avenues cut through the congested medieval Latin Quarter, displacing many of the poor who lived there, and while they greatly improved traffic circulation they also permitted the rapid deployment of soldiers in the event of an uprising. Haussmann was quite frank about this: because of his works, he claimed, 'a lot of people will be improved and rendered less disposed to revolt'. In the 1920s the architect/planner Le Corbusier found this imposed redevelopment approach thoroughly enchanting. 'My respect and admiration for Haussmann', he declared (cited in Crosby, 1973, p. 187). 'A titanic achievement — hats off!'

In the last decades of the nineteenth century a number of philanthropic industrialists established model company towns for their employees. While these were paternalistic, and one of their chief aims was to produce a satisfied and productive workforce, they also were experiments in zoning and street design. Pullman, built in Illinois between 1881 and 1885 for workers at the railway car works, had separate areas of family houses and tenements, a large public park, a central two-storey arcade with stores, a library and a theatre, and a ring railway to serve factories at the edge of the town. Port Sunlight, near Liverpool, and Bournville near Birmingham, both built in the 1890s, also had simple zoning and provided community facilities such as libraries, schools and parks. At Earswick in Yorkshire, which was planned by Raymond Unwin and Barry Parker in the 1890s, there were tree-lined roads, the first planned culs-de-sac, and every house had both a back yard and a front yard. Innovations such as these were soon to find their way into conventional planning practices.

These planning precedents came together in two separate movements at the end of the nineteenth century — the City Beautiful and the Garden City. Though focused respectively in America and in Britain, these were really parts of larger, international trends. Their proponents travelled widely, borrowing and conveying ideas as they went, for their concern, like that of many subsequent planners, was chiefly with discovering good models for planning and not with national or local character (Sutcliffe, 1981, pp. 163-201). At first they held firmly to the conviction that good, clean cities will make good people, and they argued for sweeping reforms. Such grand ideas were completely at home in the utopian atmosphere of the early century. But city planning is also a practical affair, and in order to make them practicable these idealistic principles were, as early as 1910, being reformulated into administrative guidelines equally applicable everywhere. By the 1930s it was only architects such as Le Corbusier and Frank Lloyd Wright, far removed from the day-to-day problems of city development, who continued to dream of dramatic new city forms which could solve all the problems of modern urban civilisation in one fell swoop.

City beautiful and master planning

The Columbian World's Fair of 1893 cast a long shadow on city planning as well as on architecture. The layout of the grounds with its grand vistas, superb landscaping and overall cleanliness (the manure problems of horse-drawn carriages were resolved by having students push visitors in wheelchairs) were the product of a single comprehensive plan. The result was a contrived landscape which was dramatically superior to those of real North American cities. The conclusion to be drawn was obvious: if this was what could be achieved by comprehensive planning then it clearly ought to be used to make all cities beautiful.

The city beautiful movement flourished for the first 15 years of the twentieth century, then petered out over the course of the next 15 years. Its main proponent was Daniel Burnham, the manager of the Columbian exposition. He was made responsible for directing the revival and adaptation of L'Enfant's century-old plan of malls, monuments and avenues for Washington, DC, and his plans for Cleveland and Manila were largely implemented. These all revealed the influence of his impressions of European cities, especially Rome and Haussmann's Paris, which he visited specifically to garner planning ideas. His aim was to establish 'a beauty that shall be present to do its pure and noble work among us for ever'. This was to be accomplished by realigning streets and making them into wide, tree-lined avenues that focus on to civic facilities (Figure 4.2). City halls, government buildings, theatres, libraries, and museums to record the progress of human civilisation, with status and fountains before them, all in the best classical revival styles with uniform cornice lines as used in Paris, were to front on to these grand streets.

In 1909, after more than a decade of deliberation, Burnham published his plan for Chicago. This went further than mere city beautiful planning, for it attended not only civic facilities and their appearance, but also to issues of commerce, industry and transportation, parks and the lakefront, population growth and the future course of the regional development of the city. It was the first city-scale 'master plan', setting out in detail how the city might be at some point in the future, providing a goal towards which development could proceed. Trying to mastermind changes in urban form in this way subsequently became a widespread practice, and almost every city now has to have some sort of a master plan, though since about 1950 these have been treated more as guidelines than as end results to which all development must be directed. In practice, of course, master plans set goals that are soon by-passed by social and technological changes, and which could in any case be fully realised only by totalitarian methods. Though parts of Burnham's plan for Chicago were implemented along the waterfront, and some of the avenues were built, many of his proposals were already out of date in an age of automobiles and skyscrapers and they were never adopted.

There was little concern for housing or for social reform in city beautiful and master planning. Many of Burnham's efforts were supported by clubs of businessmen and industrialists who had scant interest in such matters but who were delighted with the

Figure 4.2: City beautiful planning. A drawing for a proposed street in Burnham's 1909 Master Plan for Chicago; note the wide, tree-lined street and uniform cornice line. Northbourne Avenue in Canberra, a city planned in 1913 on city beautiful principles by a former apprentice of Burnham's

Source: Mayer and Wade, 1969, p. 279

idea of great plans for the heroic cities of industrial democracy. 'Make no little plans, for they have no power to stir men's blood', Burnham exclaimed in one of his speeches. It was a great rallying cry, yet the real effects of city beautiful were fragmentary. Certainly many North American cities have an avenue, a civic square or college campus that owe their style and layout to this movement, but in the context of the whole modern urban scene these are little more than isolated curiosities, remarkable only for their formal flower beds, and the unusually wide streets leading to clusters of classical revival buildings. For all Burnham's rhetoric this was just an aesthetic movement characterised by a sort of localised benevolent capitalist authoritarianism. After the first flush of enthusiasm city governments had neither the inclination nor the funds to carry out pretty master plans, especially while there were pressing needs for social reform, for the improvement of basic living conditions, for paving roads and for the installation of sewerage systems.

In North America the city beautiful movement flourished for about 15 years partly because cities, and especially civic centres and state capitols, were still unfinished and any imaginative ideas about how to build them were welcome. In Europe there was less opportunity for such ceremonial urban redevelopment, and in any case the Columbian Exposition had been an American and not a European event. But there are some indications of a similar beautification emphasis in European town planning at the turn of the century, perhaps following the lead of Sitte's picturesque town building, perhaps because of a continuing admiration for Haussmann's redevelopment of Paris. Thus the Kingsway redevelopment in London, carried out between 1900 and 1910, with its great processional route culminating in a crescent at India House, is certainly from a city beautification mould, albeit with the symbolism of the Empire incorporated. And *The Journal of the Town Planning Institute* was launched in 1910 claiming that its goal was 'the emancipation of all communities from the beast of ugliness' (Cherry, 1974, p. xi). Nevertheless there can be no serious question that the stronger planning influence in Europe at the beginning of the century was that of the garden city.

Garden cities

The chief responsibility for inventing the garden city goes to

Ebenezer Howard, but successful inventions rarely occur without precedent. The term itself was not original; Howard may well have taken it from the motto for Chicago — *Urbs in Horto*, the City in the Garden — where he had worked for a few years in the 1870s, and the idea of garden communities was a common victorian one, corresponding with beliefs about the need for contact with nature and the healthy emanations of trees and plants. Moreover Howard was directly influenced by Edward Bellamy's ideas about a future socialist society, and in 1893 had designed an ideal community for discussion by a club set up in England to promote Bellamy's ideas (Fishman, 1977, p. 54). In 1898 he elaborated this into a model for making new communities that would be free of most of the problems which beset Victorian cities, and published it as a pamphlet called 'Tomorrow: A Peaceful Path to Real Reform'. After some strenuous promotion by Howard this achieved considerable popularity, and in 1902 was republished as 'Garden Cities for Tomorrow'. Within a few years a prototype garden city was under construction at Letchworth, and the housing forms and layouts used there have had a widespread effect on the planning of residential areas.

The garden city was a proposal to resolve both the congestion of cities and the isolation of rural life by combining the best qualities of city and countryside in autonomous new communities, to be located at some distance from existing cities, on tracts of about 6,000 acres, with 5,000 acres for farmland and 1,000 for the town. Each garden city could therefore be almost self-sustaining. The tract of land would be purchased and held by a board of trustees who would rent parcels of it to citizens, the rent being used to pay off the capital investment and to pay for public works, so that in time all residences would be privately owned and all public facilities would be owned by the community.

The diagrams Howard drew for his pamphlet were generalised but in his written account he does indicate what he thought a garden city might look like. There would be wide, tree-lined boulevards leading to a central park of several acres and to public buildings (there may be an element of city beautiful in this arrangement); around the park there might be a glass arcade, which Howard called a 'crystal palace', filled with shops. The residential streets would be tree-lined, and the houses would be of varying styles and setbacks from the street. At least some of the houses would be served by communal gardens and kitchens. At the town fringe, as at Pullman, there were to be factories and warehouses served by a loop railway (Figure 4.3).

Howard realised that in practice the order and form of these elements would have to be adapted to the particular site. In doing this it was important that the town be planned as a whole, and he expressed this by means of a botanical analogy: 'A town, like a flower, or a tree, or an animal, should at each stage of its growth, possess unity, symmetry, completeness.' This required that there would have to be careful control over such facilities as shops to prevent over-competition and waste. It also meant that when the garden city had reached a population of 32,000 further growth should be accommodated by the development of a new autonomous garden city elsewhere. And so on, until the entire pattern of existing cities and countryside had been reconstructed.

The landscape of garden cities

As a result of the publication of Howard's pamphlet an association was formed for the purpose of building a garden city. In 1903 a site was acquired at Letchworth to the north-east of London, and development began soon thereafter. The planner and architect appointed to translate Howard's ideas into reality were Raymond Unwin and Barry Parker, fresh from designing the model village at Earswick near York. Whereas Howard was deeply influenced by Edward Bellamy's vision of the future, Unwin was a supporter of Bellamy's protagonist William Morris and had on several occasions written for Morris's magazine. This interesting difference of opinion may explain why the forward-looking concept of the garden city was realised in rustic and traditional styles.

The plans Unwin and Parker developed for Letchworth incorporated most of the features that Howard had proposed, including community facilities. Though the first houses were all conventional, the Homesgarth block built in 1909 had 32 kitchen-less units around a quadrangle, with a communal kitchen and dining room in the corner (Hayden, 1981, p. 230ff). Howard himself moved into one of these units in 1913. Similar communal developments were made at Welwyn, the second garden city which was begun in 1919, but these, and the communal playrooms and laundries and other social and political reforms of the original garden cities, never caught on popularly. In fact

Figure 4.3: Garden city planning. A theoretical segment of a garden city, and a housing layout with culs-de-sac and loops in Letchworth. It was especially the latter which influenced subsequent residential developments

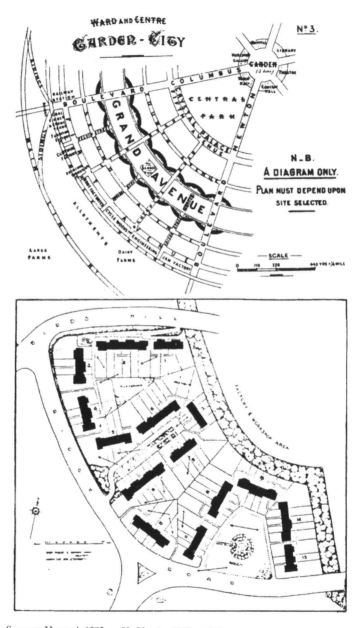

Sources: Howard, 1902, p. 53; Unwin, 1909, p. 348

Letchworth was for some years a butt of jokes and cartoons because it attracted unconventional people who were vegetarians, wore smocks and home-made sandals, and grew much of their own food.

The layout of Letchworth is carefully adapted to the site and does not bear much immediate resemblance to the diagrams of garden cities that Howard had drawn, yet most of their elements are to be found in it. There is a great boulevard park (now called the J.F. Kennedy Rose Garden) which leads to the public buildings and the railway station, there is a short glass 'crystal palace' shopping arcade, the industry is on the edge of the town, there is an agricultural green belt. It is, however, not these, but Unwin's residential layouts which have had widespread influence. In order to give variety to the residential streets and to create pleasant visual effects the houses were arranged in different patterns, in clusters around small greens, in U-shapes, and around culs-de-sac; specific site details were left to the property owners to complete but the principle of avoiding the separation of different social classes by mixing different house types was applied. Wherever possible T-intersections rather than crossroads were used, an idea adapted from Sitte (at crossroads there are 16 different ways in which it is possible to collide with another vehicle, at a T-intersection only three). These practical devices have been and still are widely used in laying out residential developments.

The architectural styles of the garden cities were a pleasant mixture of vernacular revivals (Figure 4.4). The public buildings had an updated georgian quality to them; the houses were mostly brick with red tile roofs and a rustic look. This was not a new approach to domestic architecture. As early as the 1850s William Morris had had a house designed for him by his friend Philip Webb — the Red House at Bexleyheath south of London — which rejected crass gothic revivalism and sought to revive the spirit of medieval vernacular houses. Other architects had followed this lead and had developed styles for expensive suburban dwellings derived from tudor half-timbered houses and Cotswold stone houses. In Letchworth Unwin and Parker adapted the traditional brick farmhouses and townhouses of south-east England and then scaled them down to something affordable for most of the residents. What they seem to have realised was that most people, not just the wealthy, prefer to live in houses which have some suggestion of tradition and age about

Figure 4.4: Letchworth. The neo-georgian shops and the arrangements of houses and their vaguely vernacular styles had a pronounced influence on suburban developments in Europe and North America up until the 1930s

them, and which are not all identical. This was a powerful recognition. Vernacular revivals and modifications have dominated the commercial suburban developments of Britain and North America ever since — they can be seen in fragments of fake half-timbering, in semi-georgian doorways, leaded windows, steeply pitched gables above doorways, and in decorative brickwork and stonework (Figure 4.5).

Figure 4.5: House styles from the 1920s and 1930s. a) North American bungalows, with a gable porch set inside a front gable, were built everywhere in the 1920s; b) a front-chimney style from the late 1920s, with exaggerated gable brought almost to the ground and fake half-timbering; c) an English narrow house, with a suggestion of half-timbering, about 1925; d) an English bungalow, about 1928; e) an English semi-detached house, called the Elsdale and built by the Laing company in 1937

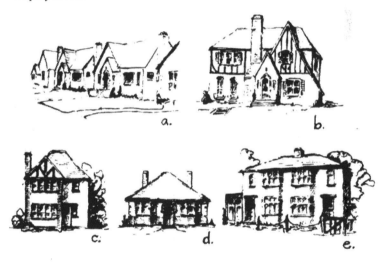

Source: redrawn from *Architectural Record*, 1932, p. 44 and 1937, p. 71, and Edwards, 1981, p. 135

While the influence of garden cities on modern landscapes has been considerable it is probably much less than both Howard and Unwin must have hoped. Certainly there are the two garden cities in England, and a few variants in America (Forest Hills and the three greenbelt towns built in the 1930s), France and Holland. There are also many new towns around the world, and thousands of 'garden suburbs' patched on to the edge of towns, which incorporate the physical features of the original garden cities. But these invariably ignore ideas about communal living and autonomy. They combine the best of the city and the best of the country only in a very limited sense. It is, in short, the planning practices, the street layouts and pleasantly rustic domestic architecture, rather than the reform ideals, which have been copied. They have, however, been extensively copied and there is scarcely a twentieth-century residential development, whether the high-rise new towns in Sweden or North American suburban sprawl, that does not owe some part of its organisation or appearance to the garden city movement, albeit modified by subsequently developed planning concepts such as neighbourhood units and land-use zoning.

Neighbourhood units

The year 1909 was an important one in the history of urban planning. Raymond Unwin's book *Town planning in practice* was published, and this promoted many of the practices of garden city planning which he had worked out at Letchworth; the first town planning act was passed in England; Burnham presented his master plan for Chicago; a national conference of city planning was held in Washington; and the first professor of town planning anywhere was appointed at Harvard University. In that one year comprehensive town planning established its legitimacy as a viable means for dealing with the problems of cities, though there were still few techniques for actually doing this other than building by-laws, the garden city approach of buying up an entire tract of land and starting from scratch, or the city beautiful method of convincing influential businessmen of the merits of a master plan. Not until after World War II was there effective legislation to enable the implementation of official plans. However, between 1910 and 1945 planning procedures were developed which have become part of the standard post-war planning repertoire and which have had a considerable effect on the appearance of cities. Most notable among these are the neighbourhood unit, the Radburn principle and zoning.

The matter of sound relationship between school and community had been a cause of concern for some time. The fact that schools stood empty in the evenings and at weekends was irksome, and there were also problems of access to distant schools for small children. About 1910 the concept of making the school with its playgrounds the central focus of a residential community had been proposed by Clarence Perry. He also argued that the furthest reasonable distance for young children to walk to school was about a quarter of a mile. The area roughly within a quarter mile radius of an elementary school therefore constituted a sort of unit, a neighbourhood unit. This model was first applied at Forest Hills, where it was found that, allowing for commercial and open spaces, and with residential densities of about 30 persons per acre, the population in a neighbourhood unit was about 5,000 (Handlin, 1979, pp. 154-7).

These ideas were revived and elaborated in the 1920s, when Clarence Perry made an important contribution to the New York Regional Plan. This plan was a comprehensive investigation of the planning issues confronting New York City; it addressed problems of population, transportation, industry, architectural and civic design, and residential development. In this context Perry was given the opportunity to develop his ideas more completely than before. He argued that the rise in automobile traffic was effectively cutting up cities into cellular blocks and that neighbourhood units were excellent ways of coping with this problem (Figure 4.6). He identified six principles which should be followed in their design:

1. Their size should be determined by the amount of housing required to support a single elementary school. The population for this was about 5,000, but the area depended on whether the people were housed in detached dwellings or apartments.
2. The boundaries should be arterial roads which would allow all through traffic to by-pass the neighbourhood.
3. Open spaces should constitute about 10 per cent of the area.
4. The school and other institutions should be at the centre.
5. Local shops should be located at the periphery of the neighbourhood unit, especially at the intersections of arterial roads where business opportunities were best.
6. The internal street system should have a varied layout with streets only as wide as was necessary for local traffic.

Articulated in this clear and practicable way, the neighbourhood unit made good sense and it subsequently became a basic

Figure 4.6: The neighbourhood unit. This is Clarence Perry's diagram in the New York Regional Plan of 1929. Other diagrams show it applied to industrial and apartment districts. The neighbourhood unit has proved to be one of the most widely adopted and enduring planning ideas of the century, and has been used as a basic organising device in both European new towns and North American suburbs

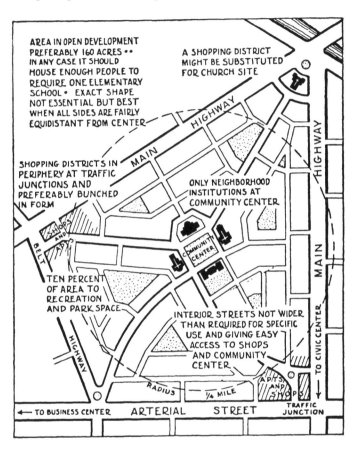

Source: Perry, 1929, p. 88

model for organising residential developments both of houses and apartments not only in America, but in Canada, Australia, Britain and several European countries. Whenever we come upon a school set in a park surrounded by houses or apartments we have probably stumbled upon one. In spite of this popularity it has not been without its problems. It creates urban islands, cut off by busy arterial roads, which may meet most domestic needs but which offer few employment opportunities. The assumption that residents would turn their attention inward has not been clearly validated; instead the centres for social activity are usually at the intersections of arterial roads and along the edges where the stores are, but these are not real centres because they are bisected by wide and busy streets. Even more fundamentally, shifts in the age structure of populations have sometimes made it necessary to close the schools, thereby undermining the essential logic for neighbourhood units.

The Radburn principle

While the neighbourhood unit was still a fresh, untarnished idea it was incorporated into the innovative development at Radburn in New Jersey. Radburn was a new town conceived in the garden city tradition (one of the main streets is Howard Avenue) and embracing all the techniques of suburban planning developed since 1900. Its great innovation was that it adapted all of these to automobiles. Its timing was, however, most unfortunate; with the collapse of the Stock Market in 1929 the company founded to finance Radburn's development lost most of its funds and the development was not completed.

Radburn was designed to cope with the facts of rising automobile ownership and the horrific number of pedestrian/ automobile accidents that were occurring in the 1920s; it was planned for the motor age. This it did by means of the 'super-block' — a form of house and street layouts that broke away from conventional grid patterns and which was partly derived from studies of garden cities in Britain (Stein, 1958, p. 44). A super-block consists of a park area rimmed by houses; the houses face on to the park and pedestrian pathways, and back on to culs-de-sac which provide automobile access to collector roads (Figure 4.7). Neighbourhood unit principles were adopted, so there is an elementary school and a playground in one of the superblock park areas; access to this from adjacent superblocks is by pedestrian tunnels under the collector roads. By all of these means a complete separation of pedestrians and automobiles is achieved. This is the so-called Radburn principle.

All that was built of Radburn were two superblocks, an apartment building and a small shopping plaza. The houses are a mixture of suburbanised American vernacular styles, mostly with

Figure 4.7: The Radburn principle. Clarence Stein's 1928 plan for Radburn in New Jersey was based on the modification of neighbourhood units into 'superblocks', essentially a park surrounded by houses which fronted on to the park and backed on to automobile closes or culs-de-sac. In this way pedestrian and vehicular traffic were separated. This principle has been extensively used in European planning, and less so in North America

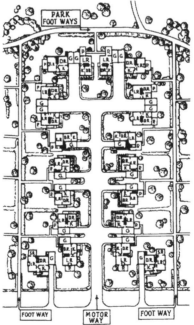

Source: Stein, 1958, p. 42 and p. 46

a colonial flavour. They have attached garages — the first houses other than those of the very wealthy to acknowledge the automobile's presence in domestic life.

Radburn was a good experiment. Its planner, Clarence Stein, has listed its merits: it was safe, orderly, spacious, and had more parkland and cost less to develop than conventional suburban layouts. Its forms have been widely admired and the Radburn principle of pedestrian-automobile separation is well known to planners. The three small greenbelt towns built in America in the 1930s under the New Deal, and post-war new towns in England and Sweden and elsewhere, all employ modified versions. Commercial developers, however, have never been comfortable with it and have not adopted it. Indeed, when Clarence Stein made the plan for the company town of Kitimat in British Columbia in the 1950s, again employing this principle, the contractors so little understood it that they simply ignored the blueprints and turned many of the houses to face the road and to back on to the park in the conventional way.

Zoning

More or less contemporary with the development of the neighbourhood unit, but wider in its impacts, was the emergence of zoning. Zoning is the practice of allocating different areas of cities for different uses, much as rooms in a house serve different functions. In Britain zoning has been adopted implicitly in the practice of fixing areas on a town plan where certain incompatible types of land use should not be permitted. In North America it is an explicit and legally enforceable means for ordering land uses, and it is the basic tool of urban planning.

Zoning by-laws were first developed in Germany and California in the late nineteenth century — in Germany to keep abattoirs out of residential areas and in California for the discriminatory purpose of restricting the locations of Chinese laundries. Between 1909 and 1915 Los Angeles extended the application of such by-laws by using them to establish a distinction between residential and industrial areas (Handlin, 1979, p. 154). The take-off for the widespread adoption of zoning came with the passage of a by-law in New York City in 1916, partly to control skyscraper development (it was the same by-law which led to the wedding cake style) and partly to proscribe commercial, retail and residential areas of the city. The city was divided into districts, in each of which only specified uses were permitted, and in each of which various building height restrictions were to be applied (Figure 4.8).

The value of zoning was immediately recognised. Writing in 1920 on the influence of the New York zoning resolution J.T. Boyd (1920, p. 193) characterised the old city as a fungus, and the new zoned city as an efficient mechanism. Zoning, he claimed, would reduce wild fluctuations in real estate values, safeguard the interests of adjacent property owners, serve to control infringe-

Figure 4.8: Part of the 1916 zoning map for Manhattan. This was not the first zoning by-law, but it was the one which stimulated the widespread adoption of zoning as a means for controlling urban land uses. The solid lines indicate streets zoned for residential or business uses, the white streets are residential only, broken lines indicate unrestricted uses

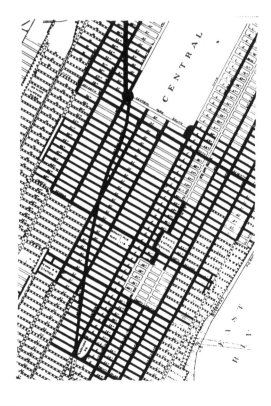

Source: M. Scott, 1969, p. 157

ments of air and sunlight rights, and organise a city into a coherent system of land use districts. In short, speculators, existing owners and city fathers all had something to gain from zoning. After their constitutionality had been upheld by the Supreme Court in 1926 zoning by-laws were passed by most American cities.

Zoning practices have become increasingly detailed and thorough since the 1920s. They have not been without problems and abuses — sometimes the standards seem to be little more than arbitrary, sometimes they have been used as an exclusionary device to prevent change and social integration, sometimes they have been used by developers and speculators to manipulate property values. Nevertheless they have had a profound impact on the appearance of cities by simply sorting out land uses. Districts set aside on maps for industry, for stores, for single-family residential uses, for apartments, for recreation, have come to manifest themselves in quite different landscapes. The detailed features of these are further defined and enforced by zoning schedules which may specify, for example, maximum permitted building heights, setbacks from the street, the width of sideyards, perhaps even building materials and styles. The result on the ground is segregated landscapes — here a zone of high-rise apartments, there a zone of detached houses, beyond a zone of retailing revealed as a plaza. And the boundaries between zones on the maps appear no less clearly as boundaries in urban landscapes — a six-foot-high fence marks the line between a residential zone and a retail zone, an arterial road separates industrial and residential uses. In older areas this sort of segregation is not so clear since remnants of uses established before zoning by-laws were passed sometimes remain (they are 'legally non-conforming uses' in the technical jargon), and owners cannot be forced to comply with zoning until the property is redeveloped. However, in those parts of cities where everything old has been carefully eradicated, zoning has directly contributed to the creation of a tight visual order in which landscapes correspond almost exactly to the land use zones set out in plans.

Unrealised ideal city plans: Radiant City and Broadacre

The roots of modern town planning lie at least in part in the ideals of the Columbian exposition, and in the utopian speculations of Edward Bellamy and Ebenezer Howard. Among planners this idealism was quickly tempered by practical concerns and translated into attempts to find the means to change actual cities for the better. Architects whose thoughts turned to planning did not feel this need to be practicable. They dreamt and have continued to dream and design for a world in which there are few historical, political or social constraints, a world in which architects would be the masters of urban form and appearance. It is difficult to say just what influence these architectural utopias have had on the course of landscape development. They are sometimes considered to be the precursors of the modern urban scene, and other architects have certainly been familiar with them and may have incorporated elements into designs for actual projects. My best guess is that the town planning proposals of architects, including the widely publicised Radiant City designs of Le Corbusier and Frank Lloyd Wright's Broadacre City, have had little impact on modern landscape development, and that any relationship between what we might see now and their drawings is largely coincidental. They have, however, been accorded considerable attention and they cannot be ignored.

In the early 1920s the Swiss architect Le Corbusier conceived the possibility of creating a totally designed modern city, and intermittently for much of the rest of his life he drew up plans for great imaginary cities or for the drastic reconstruction of existing ones. His Voisin Plan for central Paris, first elaborated in 1925, would have solved problems of congestion by the simple expedient of razing everything old to the ground and erecting a mix of low-rise terrace apartments and 60-storey towers in orderly lines. The modern city, he wrote, lives by the straight line; it is morally better than the curve, which causes cities to sink to nothing and the ruling classes to be overthrown.

The central principles of these grand plans were laid out in the declarations or manifestos Le Corbusier wrote in the 1920s about what he called the Radiant City or the Contemporary City. His dream was of a great regional scale development consisting of a central city for 500,000 people, without families, surrounded by a green belt, then a number of smaller 'garden cities' (his use of the term seems to owe nothing to Howard) for the families. The combined population would be three million. There was nothing small scale or picturesque about this proposal; everything old was to be done away with and replaced by 60-storey skyscrapers for offices and apartments, slabs of terrace apartments, and wide highways. At its heart, between the skyscrapers, would be a huge transportation facility, with roads and railways and an airport!

Le Corbusier listed the aims of the radiant city. They were:

1. To decongest city centres.
2. To increase the population density of city centres by building tall; up to 1,200 people per hectare compared with only 300 per hectare in central Paris.
3. To improve traffic circulation by replacing narrow streets with wide thoroughfares; the street, he declared, should be a traffic machine.
4. To increase open space; the tall apartments would require only about 5 per cent coverage compared with 90 per cent coverage in central Paris.
5. To provide a variety of vistas and perspectives.
6. To take advantage of mass-producing building units.

This was a magnificent dream. It was also an absolute and a totalitarian one. It is sometimes held to be the source of the modern landscape arrangements of apartment buildings regularly deployed in rows. There is certainly often a striking similarity between these and some of Le Corbusier's drawings, but the precise connections between them are obscure (Figure 4.9). It is known that Le Corbusier's ideas were adopted by CIAM (Congres Internationaux d'Architecture Moderne — a sort of club of self-produced influential architects of which Le Corbusier was a key member, and which met intermittently from 1928 to 1959). In a charter it adopted in 1933 CIAM proclaimed that 'Housing should consist of high, widely spaced apartment blocks which would liberate the necessary land surfaces for recreation, community and parking purposes.' This approach was subsequently used in many European housing developments, including the public housing at Roehampton near London in the late 1940s, though nowhere with 60-storey towers. But there are other sources for the landscape of apartment blocks. As early as the 1890s both William Morris and Montgomery Schuyler had speculated about the possibility of future blocks of apartments widely separated to permit air circulation and light penetration; in the 1920s the planners of the German design school at the Bauhaus had articulated precise planning principles for the layout of apartment buildings; and by 1930 in America some skyscraper apartments, such as Alden Park Towers near

Figure 4.9: Le Corbusian Planning. A drawing of the Radiant City showing a highway leading into the city centre with its 60-storey, cruciform plan skyscrapers. Elements of these spacious arrangements of skyscrapers can sometimes be glimpsed in the modern landscape, for instance, here on Highway 427 in Toronto, but somehow they do not live up to Le Corbusier's hopes for a new urban civilisation

Source: Le Corbusier, 1929, p. 242-3

Philadelphia, had already been built in parkland. In short, Le Corbusier cannot be understood as the single source for this landscape, though he was undoubtedly its most outspoken advocate.

Frank Lloyd Wright's ideal city, which he called Broadacre, could have been conceived in opposition to the high-rise, machine-dominated radiant city, for it was to be a low-density, mostly low-rise development in which machines were the tools of the people. 'Unless he masters it, the machine masters him' Wright had written; the machine was to make life free, happier, more joyous. Broadacre was, he hoped, the means by which this could be achieved. This was a sweeping design idea developed and redeveloped between 1921 and 1958; its central feature was a wide, democratic landscape, really no city at all, which would come to replace all the existing settlement forms of America (Figure 4.10). It was democratic because it was decentralised, with 'little farms, little homes for industry, little factories, little schools, a little university' (*Architectural Record*, 1935, p. 247). Its components were these:

— access to the natural world would everywhere be maintained, and natural materials would be used as much as possible;
— democracy was to be achieved through individual property ownership;
— decentralisation based on electrical power and new means of mobility (Wright's drawings for Broadacre include fanciful helicopters);

Figure 4.10: Frank Lloyd Wright's plan for Broadacre, a low-density settlement of small farms linked by expressways. This is sometimes claimed to be the prototype for low-density suburbs, but in reality there is probably almost no connection between them

Source: Wright, 1945, p. 55

— no commercial activity, for such activity was tawdry and the result of 'mobocracy', or mediocrity rising to high places;

— Usonia: a word play on 'US own', 'our share in the Americas'; everyone could have a homestead of a few acres if they so chose.

The idea, no less magnificent than Le Corbusier's, was that Broadacre would replace existing cities. It would be a natural city of 'organic simplicity'; there would be diverse buildings, small farms, large markets, and occasional free-standing skyscrapers. The actual construction of Broadacre was confined to a few houses in various parts of America, such as a group on Usonia Road in the exurban landscape just north of New York City.

Wright's Broadacre dream is sometimes touted as the conceptual source for low-density American surburban developments, with sprawling ranch houses on quarter acre lots, albeit without the organic and democratic ideals that Wright extolled. There may indeed be some remote connection, but there are many other far more important factors involved in such developments — rising automobile ownership, increasing affluence, the profit margin for corporate developers, planning controls. Wright's ideas are of minor significance among all of these. His was just an interesting speculation about an alternative urban form.

Comments on the invention of town planning

The concepts and methods of town planning that were developed in the first decades of the twentieth century have had a considerable impact on modern urban landscapes, though it is not the impact that the early planners hoped for. Their legacy is to be encountered in roundabouts, T-intersections, U-loops, neatly segregated land-use zones, and neighbourhood plans. But their cherished ideals and hopes for social reform and urban reconstruction have come to very little. The autonomous garden city, the Radburn principle of separating cars and pedestrians, the radiant city of skyscrapers in parkland, the decentralised city, have all been realised at best only in debased and limited forms. What appears to have happened is that these ideals were turned into models, simplified for the purpose of textbooks or classrooms or developers, adjusted to the less radical planning tools of zoning and neighbourhood units, modified by bureaucracies, adapted to political exigencies, and otherwise thoroughly watered down for ease of application and administration. Once entrenched as habits of thought these were not easy to displace or transcend, and it was in these simplified forms that they were incorporated into official planning practices after World War II. With the benefit of hindsight, and from the perspective of landscapes, it does indeed seem that urban planning has turned out to be less of a movement for social reform than a means for trying to make cities function as efficiently as factories.

5

Ordinary Landscapes of the First Machine Age: 1900-40

Throughout the first three decades of this century architects and designers searched for an aesthetic style that was suitable for machines and the buildings to house them. Eventually they decided that the unornamented, geometric, standardised styles of modernism looked right, and they then spent the next three decades trying to convince everyone else about this. In the meantime an entirely different machine-oriented landscape had already been created along city streets and country roads. It was commercial, messy, filled with poles and wires and signs and *ad hoc* architectural styles. There was nothing pretentiously modern about it; it was made by independent businessmen in response to popular demands and by municipal engineers trying to provide a safe public environment. It was a common or ordinary landscape, that is a landscape which people make for themselves or have others make for them, as a practical, ongoing thing, and which is mostly taken for granted as the background for daily life (Stilgoe, 1982). Probably what distinguishes it most of all is its lack of aesthetic self-consciousness; the municipally created parts of it, such as street surfaces and road signs, seem to have been created not to attract attention, while commercial elements like billboards call out to be noticed without regard to their context or their neighbours.

Around the turn of the century techniques of mass production were applied to durable consumer goods such as vacuum cleaners, toasters, washing machines, telephones and automobiles; the operation of these required generating stations, transmission wires and pylons, traffic signals and new types of mechanical streets and highways. The landscape which resulted was therefore an expression of more people owning and using more bought things, proudly displaying their use of them, and looking for places to buy and service them. It borrowed arbitrarily from older traditions and invented new solutions and forms as necessity demanded. This produced unlikely combinations like neo-classical filling stations and the beaux-arts traffic signals installed at Newark in New Jersey in the 1920s. Finding such impurities distasteful, both modernists and traditionalists have condemned the appearance of the ordinary landscape for its untidiness and its lack of grace or style.

Appearances can, however, be deceiving. Behind these messy ordinary landscapes there lies an almost obsessive concern for efficiency of operation and for scientific management, a concern that has permeated both business and municipal government during the last 80 years, a concern for making the world function well and not at all for how it looks.

Parkways and expressways

Although it was clear that the flourish of technological invention at the end of the nineteenth century would bring great changes to the conditions of life, it was by no means clear at the time just what those changes would be. The use of electricity was expected to lead to clean, decentralised cities but little attention was paid to the possible effects of automobiles. The advantages of practical motor vehicles such as vans and omnibuses were quickly realised, not least because they would save municipalities the great costs of removing manure and dead horses from city centres. Cars, however, were considered to be little more than recreational toys (Figure 5.1).

It was in this context that the Bronx River Parkway was conceived in 1906 as a highway for the exclusive use of cars, the first highway to be so designed. The construction of the Parkway was part of a general improvement of the Bronx River Valley in New York City in order to combat dereliction and pollution, which were not only unhealthy and unsightly but also threatened the wellbeing of the animals in the Bronx Zoo (Tunnard and Pushkarev, 1963, pp. 161-2). The highway had long flowing curves suitable for the speeds of automobiles, crossing roads were carried over it on bridges, abutting property owners had no right of access, and it was elegantly landscaped (hence 'parkway') (Figure 5.2). All this was intended for sedate recreational driving

Figure 5.1: The beginnings of the modern ordinary landscape — the automobile, the roadside advertising, and the early drive-in restaurant. A Coca-Cola advertisement from 1905

Source: Atwan *et al.*, 1979, p. 195

at a time when there were only 105,000 motor vehicles registered in America, so while it was forward looking it was scarcely a populist enterprise.

The idea of the expressway just for motor vehicles was conceived early but it took a long time coming to fruition. In fact the Bronx River Parkway was not fully opened for traffic until 1924, by which time automobile ownership in America was in the millions, Henry Ford's mass production methods were in full swing and Model T's seemed to be everywhere. By the 1920s automobile accidents and traffic congestion had become a matter of serious concern to local municipalities. In America over 30,000 people were being killed every year on the highways, and a survey taken in Manhattan in 1921 indicated that the average speed of vehicles was 11 mph, which was lower than it had been in the horse-drawn chaos of 1900. The parkway was an obvious solution not only to accidents and congestion but also for faster regional travel. Between 1923 and 1934 over 100 miles of parkways were built in the New York region, and there were similar highways built in Delaware and Detroit, though none of these had grade separations and limited access. Only in the 1930s and 1940s did all the parts begin to fall into place with the construction of expressways to the site of the 1936 World's Fair outside New York City, the building of the East Side and West Side expressways which run down either side of Manhattan, the first freeways in Los Angeles, and the functionally designed Pennsylvania Turnpike which was intended chiefly for truck traffic (Tunnard and Pushkarev, 1963, p. 167).

The fullest realisation of the new automobile highway before 1940 was, however, in Germany. In 1919 a six-mile-long, four-lane, divided speedway had been completed in the Grunewald Forest, and opened to traffic. This was an experiment that was not capitalised on until the rise to power of the National Socialists in the 1930s; between 1935 and 1941, when work was abandoned, 2,300 miles of autobahn were completed. These were real inter-urban freeways, with heavy-duty concrete pavements, and a median strip; they by-passed urban areas and were designed for speeds up to 100 miles an hour. They set the international standard for expressway and motorway construction after World War II.

Mechanical streets and the municipal landscape

By 1940 the automobile had begun to weave a concrete and asphalt web across Europe and North America. As such things go this was a relatively slow onset for landscape change — the railways had had a far greater impact in the first 40 years of their use, but then railways could not run on previously existing roads. Motor vehicles could, although they soon began to effect all sorts of detailed changes in them. Some of these are so much part of our everyday world that it is hard to imagine that they have specific origins. The altogether unremarkable asphalt road surface, for example, was introduced in the 1890s, largely because of

Figure 5.2: Early features of mechanical streets. The Bronx River Parkway in 1922, and the Fifth Avenue traffic signals, 1923

Source: Williams, 1923, *The American city*, p. 173

lobbying by organisations of bicyclists who wanted smooth surfaces to ride on; it became a necessity with the advent of automobiles because cars throw up much more dust from dirt roads than bicycles or horses and wagons.

At the beginning of the century most roads, even in cities, were little more than wide pathways, profoundly affected by weather, dusty in summer, muddy whenever it rained, and with few traffic rules or directional signs. Of the 2.1 million miles of highway in America in 1904, only 141 miles were surfaced with brick or asphalt, though in cities many streets had been paved with concrete in the previous decade (Rae, 1971, p. 31). Thereafter municipalities everywhere began to pave and widen streets into the machine spaces of the present day. By 1916 asphalt surfaces were commonplace, and with asphalt came the possibility of lane markings to separate opposing streams of traffic, an idea that was probably derived from vehicles following street cars to make their way through congested traffic. Markings were probably introduced in Michigan in 1911, special lanes for left turns came in 1914, and Stop signs were first used in Chicago in 1915 or 1916. With rapidly increasing numbers of vehicles and the ensuing congestion the need for the control of traffic flow at intersections by means of signals became urgent. The first signals were mechanical arms, known as semaphores (a word still used in some Midwestern cities), with the words 'Stop' and 'Go' written on them. These were used in Philadelphia in 1910. In the following decade the system of green, orange and red electric lights was widely adopted. These were sometimes housed in structures which were almost as much decorative as functional. An integrated system of traffic lights installed on Fifth Avenue in New York City in the early 1920s had lights at 26 intersections which were changed simultaneously by a single police officer in a 'traffic tower' equipped with three horizontally mounted lights, an ornate clock and Second Empire decorations (Figure 5.2). In 1924 General Electric perfected automatically timed lights, and then it took only a few years for most of the ornamentation to be stripped away, and the lights to be mounted vertically on a pole or hung on wires in the utilitarian fashion which has since prevailed.

By the 1920s motor vehicles were causing serious problems of parking, and municipalities began to take measures to control this. New York City and Philadelphia banned parking on their principal streets in 1922, and many other North American cities soon followed suit. During the 1920s various systems of painting lines and curbs were used to designate No Parking areas but only with the invention of parking meters did controls become effective. These were introduced in Oklahoma City in 1935, and within a year they had been adopted by 27 other American cities (McKelvey, 1968, p. 108). By 1950 almost 3,000 towns and cities had installed them. No doubt the first parking lots date from the same decades, but they are such a humble landscape feature that nobody seems to have paid much attention to them. Direction signs and road numbers were also introduced at about the same time, chiefly as a response to lobbying by motorist organisations such as the Good Roads Association whose members used their cars to range far from familiar territory and got hopelessly lost with great regularity.

So in the period from 1900 to 1930 the modern vehicle or mechanical street was first developed through a number of independent inventions and actions. These were necessary to accommodate the rapidly increasing numbers of motor vehicles but were also an expression of the growing powers of municipal governments. Of course, municipalities had long had some authority for sewerage, water supply, roads and the support of the destitute; their purview grew rapidly as increasingly technical methods were used to construct and maintain these services. At the same time their direct impact on the ordinary landscape increased considerably. Such details as garbage cans (then also called rubbish boxes, street cans or waste cans) were introduced just after 1900, and provided in large numbers by local authorities in the 1910 and 1920s as a way of trying to reduce diseases spread by flies; the designs then adopted are still common. On a larger scale, municipalities built water and sewage treatment plants, playgrounds (also introduced at the turn of the century), and swimming pools, and set standards for the design of roads, sidewalks and street lamps. Up to 1900 sidewalks were usually made of wood, but after 1910 concrete became the preferred material, possibly because of lobbying by the cement industry. Illumination standards for street lighting were established between 1900 and 1910, as electric lamps replaced gas lights. In Baltimore and other cities at least some of the overhead telephone wires were put underground, the clutter of sidewalk displays and cellar doors was removed, and the appearance of the street generally made neater.

These were all obvious accomplishments, many of which have endured. They were made possible by the new technologies of the time, made necessary above all by the pressing problems of motor vehicles, and made desirable because of the demands for civic improvement expressed by the hundreds of local societies which had sprung up in the wake of the City Beautiful and other early city planning movements and who wanted everything, the planting of street trees, the alignment of sidewalks, even the movement and behaviour of pedestrians, to be orderly. In the 1920s serious proposals were advanced, but not to my knowledge implemented, for painting lines and prohibiting stopping on sidewalks in order to regulate pedestrian flow. In this flush of expansionism attempts were also made to impose moral standards (this was the era of prohibition, after all), and in many communities there were by-laws restricting the wearing of bathing suits except at designated beaches and swimming pools; citizens were required to wear a coat over their swim suit as they drove their Model T to the publicly maintained beach along the new mechanical streets.

From the perspective of urban landscape what is striking about all these municipal improvements in the look of streets is how visually unremarkable they were. It is as though the major features of the mechanical street were designed not to attract attention but just to be a grey background of poles, surfaces and innocuous direction signs — a functional context for an efficient city. Though it was probably not what was intended, against this neutral backdrop exotic commercial buildings and signs acquired a dramatic prominence.

Early commercial strips and the decline of main street

Automobiles required not just new roads but also facilities to sell, to store and to service them. One of the first reinforced concrete buildings in the world was the Garage Ponthieu in Paris, designed by Auguste Perret in 1905 in a functional style with something that could be an oversized radiator ornament above the entrance. Garages and filling stations were, however, rarely given such special design attention until the 1920s. Mostly they were fitted into converted barns, stables or warehouses. The first single-function gasoline stations date from just before World War I. At about the same time architects began to provide garages rather than stables for the more expensive houses which they designed.

83

But it was in the 1920s that the major commercial impact of automobiles on the roadsides of the world occurred. With soaring car ownership filling stations were required everywhere. They came in all sizes and looks, to the dismay of many architects. A.G. Guth, writing in the *Architectural Forum* in 1926, criticised the 'Chinese Pagodas, Mohameddan Mosques, Norman Castles and Flemish Towers' which he saw everywhere, and argued that gas stations should be built of local materials in traditional and conservative styles. The service stations along the Bronx River Parkway must have pleased him, for they were pleasantly rustic affairs of stone and wood shingles. By the late 1920s, however, cars were more than recreational toys. They were necessities of life for many people. Service stations were being planned as the centres of future community development, and were competing for business by means of whatever eye-catching design their owners could imagine (Lonberg-Holm, 1930). They took corner lots, the most visible and the most profitable sites available, and erected flashing electrical signs to catch the attention of drivers both by day and by night.

Travelling at 30 miles per hour in a car one sees much less than on horseback or on foot. Speed blurs details, signs have to be big and bright, land uses can be mixed up and spread out because distance is of no great importance to a driver. The result is the all-too-familiar casual chaos of commercial highway strip sprawl and ribbon development. It was not immediately apparent that this was going to happen. In 1923 the state of Indiana removed more than a million advertising signs from about 4,000 miles of state highway because of the problems they were posing for drivers' visibility and for weed cutting. The state received considerable co-operation in this, especially from larger firms who took the view that 'in this age of greatly augmented motor traffic and higher speed of travel, roadside advertising is obsolete' (Williams, 1923, pp. 484-5). Presumably they thought that everyone was going too fast to see the signs. This opinion did not long prevail.

What did happen was that motor traffic brought in its wake not only the mechanical street but also all the dross and glitter of commercialism. In the early 1930s a survey of a 47-mile stretch of highway from Newark to Trenton in New Jersey counted 300 gasoline stations, 472 billboards, 440 commercial uses and 165 intersections (Tunnard and Pushkarev, 1963, p. 162). It does not take much imagination to realise what this must have looked like, especially with numerous smaller signs and a multitude of different building styles. Appearance was, however, of much less importance than a good entrepreneurial possibility. Here were all sorts of opportunities for quick profit-taking and for discovering economic niches. New commercial facilities, specifically catering to drivers, were created and added to the confusion. Drive-in restaurants, drive-in cinemas, the first cloverleaf intersection at Woodbridge in New Jersey, the first auto camps (also called tourist camps and, after about 1940, motels), the first shopping plazas, all date from the late 1920s and early 1930s.

Many of these facilities may have been a response to the problems of parking. Main streets in the old manner provided limited parking and were ill-suited to the new mobile life-style. Anyway, by 1940 these difficulties had been sidestepped or resolved in favour of mobility, and a correspondent to the *Architectural Record* (Vol. 87, June, p. 101) could declare unequivocally that 'Today's housewife drives to her neighbourhood shopping center' because by then such automobile-oriented shopping centres were commonplace. In the late 1920s concessions had been made to cars by setting new stores back from the street line to permit off-street parking in front of them. Thus the neighbourhood plaza in Radburn, a development explicitly designed in 1928 for the 'Motor Age', has a row of spaces for angle parking in front of it (it now also has parking behind on land that was originally intended to be a small park). The idea of drive-to shopping caught on quickly and by 1932 the basic plaza arrangements of a L-plan at a corner site and a U-plan at mid-block, both with the parking in front and directly accessible from the street, had been conceived (Figure 5.3). Forty years later plazas were still being built in much the same forms, just bigger.

In 1935 the *Architectural Record* held a major design competition on the theme 'Modernize Main Street'. Competitors had to submit designs for drug, clothing and food stores and a service station. The award winners were spare and modernistic, with surfaces of shining black vitrolite, chrome fittings and modern neon signs. They were all splendid, except that by 1935 Main Street was obsolete. It was an inheritance from the horse-drawn days of the nineteenth century and before, and though existing main streets could be updated they were not well suited to the requirements of automobiles. In the proliferating neighbourhood units of the suburbs main streets had no role to play and were no longer being built, instead the parking lots of the plazas were

Figure 5.3: The decline of main street and the first plazas, about 1932

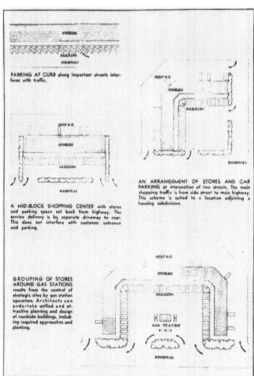

Source: *Architectural Record*, 1932, pp. 325-7

increasing in size with every new development.

Of the designs submitted only those for service stations came close to grasping the character of the new automobile-oriented landscape. The winners among these were in the International Style then so fashionable in Europe, but then the competitors could follow the example of the Standard Oil filling stations which had been introduced in 1931, and which were simple glass and metal boxes, unornamented, designed for a machine age. Other oil companies wishing to establish a progressive image soon turned to similar styles and abandoned the commercial eclecticism that had prevailed during the previous decade. In 1937 Texaco introduced a line of standardised service stations which have not changed significantly to this day. They had models for a corner lot, an interior lot and a highway location — each one clearly identified by Texaco's symbol (Figure 5.4). It was important for the travelling public to be able to find their favourite brand of gasoline no matter where they went, so all service stations belonging to one company had to have the same basic look. Corporate competition was replacing rampant individualism as a feature of the ordinary landscape. A similar process was also beginning in the food and hotel businesses with the spread within regions of America of such chains as White Castle restaurants and Howard Johnson's hotels, both using standardised, easily recognised designs. Such standardisation only became fully developed in the 1960s and 1970s, when it turned out that it was not so much undoing the visual disorder of commercial strips as replicating the same sorts of disorder everywhere. By then both main streets and the exotic gas stations of the 1920s had become so antiquated that they were not infrequently the objects of heritage preservation efforts.

By 1939 the automobile had become the primary force in determining the appearance of the ordinary landscape of cities. All that remained to be constructed was a great symbol to the foremost machine of what Reyner Banham has called the First Machine Age, the period between about 1900 and 1940 when machines became domesticated companions of daily life. Just such a symbol was conceived by the reputable French engineer Eugene Freysinnet for the 1937 World's Fair in Paris. He called it *Le Phare du Monde*, 'the lighthouse of the world'; it was to be a 2,300 feet tall tower you could drive up by means of a spiral track, and at the top would be a restaurant, hotel, solarium and garage for 400 cars (Figure 5.5). To drive up and down would

Figure 5.4: From idiosyncratic individualism to corporate standardisation — gas stations in the 1920s and 1930s. The exotic styles of the 1920s, here represented by a small castle still occupied in Lawnside in Chicago, and a medievalish station at Radburn, gave way to International Style designs like those of Texaco in the 1930s

Source: *Architectural Record*, September 1937, p. 71

Figure 5.5: Le Phare du Monde. *Eugene Freysinnet's proposal for a monumental drive-to-the-top tower for the Paris Exhibition of 1937*

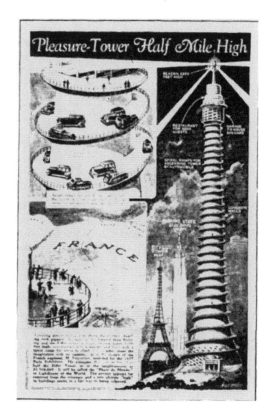

Source: *Architectural Record,* 1934, p. 41

surely have been a symbolic act, a twentieth-century version of walking a spiral maze in the Middle Ages, except of course that instead of spiritual enlightenment one would have probably received a bumper sticker saying 'This car climbed *La Phare due Monde*'. It was never built.

Conspicuous mass consumption

The term 'conspicuous consumption' was invented by the American economist Thorstein Veblen in 1899 to describe the behaviour of a new wealthy class in displaying their affluence. 'In order to gain and to hold the esteem of men,' he wrote (p. 38), 'it is not sufficient merely to possess wealth or power. The wealth or power must be put in evidence, for esteem is awarded only on evidence.' The most obvious way of putting wealth in evidence is to be ostentatious in one's acquisition of goods and in one's non-productive consumption of time, that is in one's leisure. This could be done through building excessively lavish houses, as the Rockefellers and Vanderbilts did, by employing countless flunkies and servants, holding lavish balls and by lounging around on yachts; on a smaller scale it was done by driving around in luxurious, hand-built automobiles.

The development of mass production techniques changed all this by bringing previously unaffordable goods within the financial range of large segments of the population. Status and conspicuousness, hitherto reserved for the few, now became possible in a modified form for the many; conspicuous consumption became conspicuous mass consumption. The car was the lead object in this because it was expensive enough to be a status symbol, but inexpensive enough to be affordable. Movies and movie theatres too were significant, because they made glamour and romance vicariously accessible for everyone. Houses became the object of displays of new bourgeois affluence, not on a grand scale but at least sufficiently to show that one was no worse off than one's neighbours. Perhaps this sort of social competition has always existed, but in the twentieth century with mass produced goods and with substantial increases in disposable income it has manifest itself on an unprecedented scale.

The sort of individualistic commercialism that developed in the 1920s did not meet with the approval of architects and planners. If they mentioned it at all it was to condemn it. For example Raymond Unwin disliked all commercial display and he hoped through his planning efforts to promote 'the gentler and finer instincts of man' (Creese, 1967, p. 16). Presumably that meant the sort of traditional, rusticated charms that came with the partial self-sufficiency he had designed for the residents at Letchworth. On the other side of the Atlantic Frank Lloyd Wright expressed the same distaste for commercial enterprises. 'Canned Poetry, Canned Music, Canned Architecture, Canned Recreation', he wrote in 1927 (p. 395). 'All canned by the Machine.' He claimed that he was concerned with beauty and truth, and hoped with his design for Broadacre to 'conquer sordid ugly commercialism' (1958, p. 11). The European modernists found it no easier to come to terms with the untidy gloss of advertising signs and automobile landscapes, and professed a strict preference for tidy white well-ordered buildings and land-scapes, though Le Corbusier did have one deep lapse when in 1930 he designed for Nestlé a movable fair pavilion that was veritably covered in signs (Figure 5.6).

Figure 5.6: Le Corbusier's exotically decorated Nestlé pavilion for a trade fair in Stockholm, 1930

Source: *Architectural Record,* 1930, p. 6

For some the new consumers' landscape was considered to be insufficiently modernist, and for others it was an effrontery to tra-ditional standards. Such criticisms were probably of little conse-quence to the corporations promoting their products, to the businessmen who sold them, or to the bulk of the population who wanted to buy and use them. They were variously concerned with making a profit, or making a living, or with their own satis-faction. Nevertheless the treatment of landscapes in the interests of profit and material satisfaction became so cavalier that some criticisms were certainly warranted.

Perhaps the most strident voice of protest was that of Clough Williams-Ellis, a British architectural critic who, in a book called *England and the octopus* which was published in 1928, took issue with many of the landscape developments that had occurred in

England since 1900. Though this book was directed chiefly at inter-war urban sprawl he managed to find fault with almost every new type of building or land use feature. His final thrust was a 'Devil's Dictionary' — an appendix in which he listed his greatest dislikes. This is worth reproducing here in an abbreviated form since it conveys many of the problems of the inter-war ordinary landscape of Britain. The equivalent landscape in America varied in detail and scale, but an American dictionary would not have been very different. The entries include:

Advertisements — especially billboards.

Aerodromes — their numbers and their appearance (in the US in 1929 there were already 1,330 airfields).

Archaeologists and Antiquaries — 'Their gimlet eyes have a special and peculiar focus'. Williams-Ellis disliked preserved relics.

Assimilation — the lack of it in so many commercial buildings — such as bright pink cafés which glared even at a distance.

Broadcasting — especially the fields of transmission towers then required.

Borough Engineers — for using manufacturers' catalogues and promoting mindless uniformity.

Bungalows — especially in sporadic developments.

Electric Power Distribution — which was done by locating pylons without aesthetic sensitivity.

Golf Courses — specialised parks for the few.

Oxford and Cambridge — for their intellectual insensitivity.

Paint — because of the garish new colours.

Railings — too many of them, fencing things off.

Soldiers — where the Army and the Navy settle there is unpleasantness.

Standardisation — not because it is inherently wrong but because it is so often used improperly.

Water — too frequently filled with junk and ignored.

From Williams-Ellis' list what emerges is a picture of an English landscape in a state of tension (Figure 5.7). The new technologies of automobiles, electricity, aircraft, concrete, steel and glass were confronting the well-established landscape forms of earlier centuries. Streets designed for horses and pedestrians were not easily adapted to trucks and cars. The tried and true

Figure 5.7: Roadside signs, poles and flags surround a tearoom in England in the late 1920s

Source: Williams-Ellis, 1928, endplates

styles of the gothic and the renaissance did not seem appropriate for mass-produced goods, though it was still unclear just what styles were appropriate. The traditional standards of taste and fashion no longer applied. Cities were growing at an unprecedented rate, cars and buses were conveying people in great numbers everywhere, work had been separated from home, suburbs were expanding and commercial ribbon development was reaching along the roads. The new technologies were certainly changing things dramatically, and from all the appearances of the ordinary landscape they were quite out of control. This was, in fact, not the case.

Invidious scientific management

Many of the changes to the ordinary landscape in the early part of the century now seem unexceptional, even obvious. Less obvious, and certainly not disorderly, was the economic process which lay behind them, a process which is at the heart of twentieth-century society and its landscapes — the scientific management of industry and commerce.

The principles of scientific management by F.W. Taylor was published in 1911. This book was a clear statement of the principles that Taylor had been advocating and applying for some years, principles of efficiency in the workplace as determined by time and motion studies. Using these methods Taylor had more than doubled productivity in the section of the Bethlehem Steel Works where he was employed for three years at the turn of the century.

Taylor was not a relaxed individual. He was driven by some sort of compulsion to reduce everything, including the sports he played in a fiercely competitive manner, to its mechanical components and actions. These he then rearranged to maximise efficiency. At tennis his backhand was poor so he fashioned a spoon-shaped racquet which sufficiently improved his play to enable him to win the US doubles championship in 1881 (he also invented the winch for tightening the net and the double netting) (Andrews, 1964, pp. 62, 68). His object was simply to achieve results. This determined attitude he applied to industrial processes by watching employees carrying out tasks, dividing these tasks into parts, timing each action, and devising better sequences of actions to achieve increased output. The workers' tasks could then be rearranged and they could be set to compete against the clock, because the sole purpose of labour was to produce as many of something as efficiently as possible. This was not quite the simple exploitation of workers it appears to be. Taylor's argument was that increased productivity would lead both to greater prosperity for the employer and to increased leisure time for the employee; furthermore managers, undertaking the time and motion studies and administering the courses of action they recommended, would have more to do than before and their status would grow as a result, a prediction that has certainly been realised in the second half of the century.

It would probably be a mistake to make too much of Taylor's individual responsibility for the changes in industrial management practices that followed upon his invention of time and motion study. He was a man of his times and his times were marked by a widespread concern for the reduction of waste and inefficiency. He was, however, the individual who promoted the new management approach most effectively. His book on scientific management was received enthusiastically, and it was published in several languages. Any doubts there might have been about ideas of time and motion and productivity were effectively scotched by World War I, when demands for maximum output were paramount, restrictive practices were treasonable, and many efficiency experts from industry were moved to government departments. Thereafter efficient management and centralised administration became essential goals of modern business and government.

Scientific management is not as removed from the early consumer landscape as it might seem. It was instrumental in increasing the availability of consumer goods, and in increasing workers' incomes and their leisure time. One of the earliest large-scale implementations of Taylor's scientific management approaches was in Henry Ford's 1913 assemble line for Model T cars. What Ford and his managerial assistants achieved was, first of all, the identification of a potential popular demand for a cheap and relatively reliable automobile. To meet that demand they reorganised the manufacturing process, specifically by subdividing the tasks, standardising parts and procedures, and separating the actual manufacturing process from planning and management. The result was a highly co-ordinated assembly line; the parts were brought to the chassis as it moved along a conveyor, each worker carried out a specific duty, and at the end the finished car rolled down a ramp and was started by engaging the gears. The cars were all black because that was the only paint that would dry quickly enough to permit the efficient operation of the assembly line. Since over five million Model Ts were made between 1908 and 1927, when production was stopped, the black Model T became in itself a major element of the ordinary landscape in America. In addition, it and its competitors and equivalents in other countries were the primary cause of the new highways and highway commercial facilities which were being constructed.

Taylor realised that higher wages might be necessary to induce workers to accept scientific management; they were understandably less than enthusiastic about repetitive work and overseers with stop watches and notebooks. And indeed Ford found that he had to offer high wages to reduce a turnover rate of employees which in 1913 reached 360 per cent, so he paid $5 a day to assembly line workers, twice the going rate, on condition that the workers work first for six months at the factory. With the increased productivity achieved by assembly line methods Ford could pay such wages and still reduce the price of the Model T to below $600 so that its sales could continue unabated.

Similar developments in other areas of industrial production had similar consequences. The overall result was that productivity went up, wages went up, working hours came down; more people had more money to spend, more things to spend it on and more time to spend it. (In Britain the process was slower than in America, and substantial leisure time for workers was achieved only with the passage of the Holidays with Pay Act in 1938.) By 1913, when Ford's assembly line came into full production, Thorstein Veblen's theories about conspicuous consumption and the leisure class were already dated. His argument had been that great wealth was most dramatically displayed in ostentatiously non-productive activities because leisure was such a precious commodity. With scientific management and the growing power of unions, there had been a reduction in working hours, leisure had lost its scarcity value and every class became in some measure a leisure class and a consuming class.

There is an age old tradition in which the poor and under-privileged copy, in cheaper forms, the fashions established by the wealthy and the powerful. So the fashion of conspicuous consumption was adapted to the means of the lower classes and the goods available to them. No doubt this involved great material satisfaction, but it was achieved at a considerable social price because scientific management, by isolating individual workers to do repetitive tasks, helped to separate the last of the webs of mutual aid and co-operation which had once existed in workshops. The needs for community involvement now had to be satisfied vicariously through the acquisition and display of goods. In all of this mass consumption was the keyword, and it was mass consumption which filled the towns with paved highways, cars and gasoline stations, and the countryside with advertisements for new improved brands of engine oil.

Mass consumption requires, of course, mass production, and that is best undertaken by large corporations. Those companies, such as Ford, which were able to respond to the possibilities offered by a mass market by using the principles of scientific management, were enormously profitable and expanded rapidly. Those which failed to respond either went under or were taken over. By these processes a new economic order was establishing itself. In 1923 Thorstein Veblen wrote a study of absentee ownership and business enterprise in which he identified the character of this new order. It was, he wrote, no longer the product and quality of workmanship that were at the heart of economics, but the making of money and corporate profit (Figure 5.8). That was to be done by two means, paradoxically almost opposite in their manifestations. On the one hand there were 'the colorless and impersonal channels of corporation management' (p. 215), ruthlessly efficient and objective in their drive to maximise productivity and profit. On the other hand was 'competitive salesmanship', based in bargaining, effrontery, make-believe, packaging, advertising, images, and 'salesmanship in the place of workmanship' (pp. 107, 78).

Figure 5.8: The frieze on the Toronto Stock Exchange depicting the merits of the regimented organisation of industry advocated by F. W. Taylor, the sort of organisation and management which stood behind many of the ordinary landscapes of the 1920s and 1930s

The ordinary commercial landscapes of by-pass suburbs and ribbon developments, commercial strips and plazas, were largely a result of competitive salesmanship. It was in and through these that most of the money was to be made, and few people had either the time or the inclination to worry about their visible coherence. At the same time the impersonal channels of corporate management were creating for themselves a special colourless landscape of skyscraper offices. These two landscapes may look very different; they were, and still are, two sides of a single coin.

6

Modernism and Internationalism in Architecture: 1900-40

The efforts of architects to find a style they considered appropriate for mass-produced consumer machines and their buildings eventually came to fruition in the 1920s. The style they invented drew on several earlier experiments and theories, but it had no clear historical precedent and it seemed to be as original and as profound as classical and gothic styles. It consisted of undecorated surfaces, clear edges, mass-produced components, angular shapes and the honest expression of materials. It was for a while referred to as 'The International Style'; it is now more frequently called simply 'modernism'.

The inventors of modernism were confident that they had conceived a bright, white, rational architecture that would not merely be suitable for but would actively promote the creation of a forward-looking and rational society. This would be a society completely at ease with the new technologies and their products, a society that would not copy the mistakes of history. Modernism would be the new architectural style for a new social order.

It was not to work out quite like that. Modernism never proved very popular as a style for houses, and by the 1980s its originality and acceptability even as a style for institutional and corporate buildings had begun to decline sharply as it faced a challenge from the eclectic revivals of 'post-modernism'. Nevertheless, modernism was for several decades the most acceptable style for corporate, institutional and apartment buildings, and as a result it has come to dominate the skylines and centres of cities everywhere. Clusters of apartment buildings in Amsterdam and Ulan Bator, skyscraper offices in Sydney and Singapore, boxy factories, angular phone booths and bus shelters everywhere, are all its progeny.

Sources of modernism

Modernism seems to have originated around the turn of the century in several independent reactions against pointless ornament on manufactured goods, and against stale revivals of gothic and classical architecture. William Morris, for instance, pleaded for simplicity of design, a cleaning out of all sham, waste and self-indulgence, especially in decorative arts. He thought that this might be achieved by combining the skills of artists and craftsmen to create a simple yet popular art; practising as he preached he designed and produced wallpapers and fabrics and established a hand printing press. These practical demonstrations soon acquired a following known as the Arts and Crafts Movement which sought to revive decorative honesty and sound workmanship in manufacturing and building. It was both popular and influential. Raymond Unwin and Barry Parker, the planner and architect of the first garden cities, were part of it; in America the movement was centred around a magazine, *The Craftsman*, which flourished during the first 15 years of the century; and in Europe Morris's ideas held considerable sway up to the early 1920s, when the Bauhaus, the German design school which is at the very heart of modernism, incorporated them into its initial manifesto.

A contemporary development was that of Art Nouveau (also called Jugendstil). This artistic movement of the 1890s was significant because it constituted the first clearly different alternative to long-established design traditions. It emphasised sensuous, organic, curvaceous shapes, and elongated lines that might have been suggested by the curves of the Eiffel Tower. This style was swiftly adopted in illustration, for instance by Aubrey Beardsley, but in city landscapes its manifestations were mostly restricted to detailed decorations on store fronts and entrances to some of the Metro stations in Paris.

In North America Art Nouveau had even less impact, though Nicholas Pevsner has argued that Louis Sullivan was influenced by it because of the intricate, decorative surfaces he designed for his buildings. In the 1890s Frank Lloyd Wright was an apprentice in Sullivan's office and he, too, had a fascination for nature and organic forms. Wright was to become one of the most consistently original architects of the twentieth century. Perhaps this originality was partly because he had a limited formal education in architecture, just two years in an engineering programme at the

University of Wisconsin, and had not been steeped in design conventions. Certainly the houses he designed at the turn of the century were very different from the Queen Anne and other eclectic styles then in fashion. They have a striking geometry of horizontal planes, cubes, strips of windows, and low-pitched roofs with wide eaves. Long and low, fitting into their site, these came to be known as prairie style houses (Figure 6.1). Though they were widely admired and publicised in magazines like the *Ladies Home Journal*, they did not prove easy to copy. Indeed Wright's work in general defies copying, and a clearly defined school of Wright-style architecture never developed. However, when a portfolio of his designs was published in Germany in 1910, with its illustrations of buildings that seemed to owe nothing to tradition, this was enthusiastically received by forward-looking European architects and was a major source of inspiration for them.

Figure 6.1: Frank Lloyd Wright's Robie House, Chicago. With wide eaves and a strong horizontal emphasis Wright's prairie style houses were dramatically different from the revivalist styles of the nineteenth century. The Robie House was filled with innovations — a slab foundation, an integral garage, indirect lighting, above grade laundry room, and astronomically calculated overhangs to get maximum shade in summer and sunlight in winter

The first decade of the twentieth century was a time of remarkable invention, imagination and intellectual upheaval. At about the time that Wright was developing his original architectural designs in America, physics in particular and science in general were being profoundly reoriented by Einstein's formulation of special relativity theory. In art an equally revolutionary change was occurring — the break from representational art and the invention of abstraction. In 1907 Picasso painted 'Les Demoiselles d'Avignon' in what came to be called a cubist style, and in that and the following year he and Georges Bracque painted a number of landscapes, portraits and still-lifes in the same angular, lumpy manner. The idea of paintings as necessarily having to represent objects or scenes from a single viewpoint, an idea which had held sway for centuries, had been abruptly undermined. From then on artists were free to experiment with ways of conveying sensation, experience, speed, and several different perspectives simultaneously in whorls of colour, line, shape and abstraction.

These changes in art had swift effects on contemporary architecture. Both Hitchcock and Banham suggest that the unornamented and angular forms of modern buildings, and the fact that they have to be appreciated from many perspectives and as one moves around and through them, follow very closely the principles laid out by the first abstract artists. As the artists moved to abstraction so the architects developed functional engineering styles of building; as the artists developed their representations of speed and multiple perspective so the architects translated these into built forms which have no preferred façades, no fronts and backs. Robert Hughes argues (1980, p. 12) that abstract cubist art was a response to experiences of constantly altering landscapes as seen from a train or automobile, 'a succession and superimposition of views, an unfolding and flickering of surfaces'. If he is right then these artists must be understood as a bridge between the new technologies and the realisation of modernism in landscapes. The full sequence would be this: first the new landscape experience, then the artists' representations, then the architects' constructions, and finally the making of a new landscape conforming with machine-age experiences.

Abstract art has never enjoyed widespread popularity. This is significant, because if it really has influenced the character of architecture then some understanding of it is needed in order to appreciate modern buildings and landscapes. Judging by the vaguely vernacular houses most people select for themselves and by the mass-produced paintings on living-room walls, the eyes of almost entire generations date from an era long before their own birth and are fixed firmly in the single perspective representations which were the only ones known to their late-victorian great-grandparents. The modernist abstract aesthetic may be the one by which new urban landscapes have been built, yet hardly any-one likes it or understands it.

Functional futurism

The different sources of modernism — the Arts and Crafts Move-ment, Art Nouveau, Wright's new architecture, and abstract art — had in common a rejection of traditional styles. This rejection came to focus with increasing clarity on the condemnation of use-less ornament, and nobody went further in this condemnation than the Austrian architect Adolf Loos. In an essay on 'Orna-ment and Crime' written in 1908, he declared unequivocally that 'the evolution of culture marches with the elimination of orna-ment from useful objects' (cited in Banham, 1960, p. 94). The implication was, as Banham (1960, p. 97) has noted, that archi-tects should forthwith build like engineers in a manner suitable for a machine age. And indeed Loos designed and built a house in Vienna in 1910 to prove this very point, a house so spare and free of the dreadful crime of ornament that it anticipated the great changes that were about to occur in architectural styles by about 20 years (Figure 6.2).

Although he was part of a group of like-minded artists and writers known as the Secession, this building by Loos was very much an individual statement. Modern unornamented styles received a more practicable if less dramatic demonstration in the work of the Deutscher Werkbund, a loose organisation of German artists and craft firms founded in 1906 to advocate mass production and to find suitable designs for mass produced goods. The key figure in this group was Peter Behrens, who was architect-designer for the German electrical corporation AEG. For their buildings Behrens developed a sound industrial style, with strong forms, no trivial decoration, and a down-to-earth practical appearance for industry and machines. He was also sen-sitive to the needs of the workers, and incorporated high standards of ventilation and illumination into his buildings.

Figure 6.2: The Steiner House by Adolf Loos, Vienna, 1910, and the Fagus Factory by Walter Gropius and Adolf Meyer, 1911, anticipated and influenced the modernist styles developed in the 1920s. Such styles would have been inconceivable only ten years before, yet they would still look up-to-date if they were built now

Other members of the Werkbund followed his lead. In particular a small factory for the Fagus Shoe Last Company (lasts are the forms upon which shoes are made), designed in 1911 by Walter Gropius who was to become one of the most influential architects of the modern movement, in conjunction with Adolf Meyer, became a model for industrial buildings for the next 50 years. The interior was designed with careful consideration for good working conditions. The exterior, an undecorated rectangular form made partly of brick and partly of the new materials of glass and steel, presented a functional and no-nonsense face to the future.

At the same time the Fagus factory was being designed and built a group of Italian artists was energetically formulating an ideology that extolled everything modern. They called themselves Futurists. Their spokesman, Filippo Marinetti, wrote a manifesto in 1909 in which he ridiculed everything old, and praised machines, electricity, light, speed, factories, aircraft, great bridges and the cleansing qualities of war. 'We establish Futurism', he exclaimed (1909, pp. 42-3), 'because we want to free this land from its smelly gangrene of professors, archaeologists ... and

103

antiquarians ... Take up your pickaxes, your axes and hammers and wreck, wreck the venerable cities, pitilessly!' Since the Renaissance people had looked to former ages for inspiration and for confirmation of the meaning and value of whatever they did. Marinetti and his fellow Futurists made explicit and absolute what had already been hesitantly expressed by Edward Bellamy and many others — no more looking backwards, from now on only look forward, to the future.

There was one architect associated with the Futurists — Antonio Sant'Elia. He died before any of his designs could be realised but he did make many dramatic drawings of future cities; they showed monumental buildings of concrete with bold forms and huge parapets, towers with illuminated advertising signs at their summit, great generating stations and airfields. These futurist cities seemed to be striving to glean every possible benefit from recent advances in science and technology, translating them into original built-forms (Figure 6.3). It was, in fact, not until the 1950s and 1960s that elements of Sant'Elia's city-

Figure 6.3: One of Sant'Elia's futurist drawings for La Citta Nuova, 1914, perhaps realised in part in the forms of Scarborough College, Toronto, built in 1968

Source: Banham, 1960, p. 109

scapes were actually created in great poured concrete megastructures and institutional buildings such as Scarborough College in Toronto (within the massive grey concrete walls of which this is being written). The Futurists, it seems, had broken so radically with the past that they had far more in common with the 1980s than with the world of 1900.

An interlude — World War I

Then came World War I, the first machine war, the first modern war, a war which was by almost any measure a dramatic event in Western history. The ancient empires of Europe collapsed, a new social order arose, millions died and the war's horrors insinuated themselves into the poetry, literature and collective memory of Western civilisation. Such a war must surely have had a profound impact on the course of modern architecture and landscape development. But we look for it almost in vain.

It is conventionally argued that the man-made landscape acts as a form of historical record of major social changes. For instance, the new social orders created by the industrial revolution were made manifest in the factories, undifferentiated rows of workers' houses and the overdecorated villas of the *nouveau riche*. But World War I, though a great gash in modern history, did not significantly alter the way the world looks. Of course there are local signs of what happened. Paul Fussell, for example, in his book on *The Great War and modern memory*, writes (1975, pp. 69-70): 'Today the Somme is a peaceful but sullen place, unforgetting and unforgiving ... To wander now over the fields destined to extrude their rusty fragments for centuries is to appreciate in a most intimate way the permanent reverberations of July 1916. When the air is damp you can smell rusted iron everywhere, even though you see only wheat and barley.' Perhaps the eye of imagination is involved even in this description. It is in fact easy to drive through the old battlefields on new roads and to see almost no signs of them except for a glimpse of a military cemetery. Many of the devastated towns of Belgium and northern France were rebuilt as copies of themselves as they had been before 1914, presumably in a deliberate symbolic effort to undo the reality of destruction. And there are numerous new houses, autoroutes, motels and shops which simply cover up any remaining evidence of the Great War.

I do not doubt Fussell's argument that the war deeply changed European consciousness. But landscapes are records of construction rather than destruction, and, except for appropriate monuments and memorials, evidence of disaster and suffering, even a disaster as great as World War I, is removed as quickly as possible. Perhaps this is why planners and architects seem deliberately to ignore wars and their effects. Raymond Unwin, the English planner, once declared that he did not allow himself 'to be distracted by the immediate and distressing alarms of political, military and economic crises' (cited in Creese, 1967, p. 9). It is difficult to know how to react to this, whether to condemn it for its callousness or to praise it for its faith in the long-term continuity of social life and civilisation. I may not like its indifference, but I also have to recognise that landscape changes occur in long, slow cycles. For all the deaths, suffering and profound social changes it caused, the Great War was simply too brief to change the course of modern landscape development. From the perspective of architectural and planning history it was but a brief interlude in a continuous development from the end of the nineteenth century to the present, and its only major effects were to promote scientific management in business and government and to hasten the onset of modernism.

The Bauhaus

At the fulcrum of all histories of modern architecture and landscape is the Bauhaus. A school of design (and subsequently a building of the same name) the Bauhaus flourished in the 1920s, drew on all the strands of thought about modern, functional, futurist design for mass production that had developed in the previous 30 years, and wove them together into the distinctive style of the twentieth century. There were contemporary and similar developments elsewhere, notably in the work of Le Corbusier and some Dutch artists and architects, but it was the Bauhaus which had the most immediate and the most widespread impact on how the world looks. The appearance of buildings, chairs, fabrics, light fixtures, kitchens, desks, city skylines of angular skyscrapers, indeed almost anything that we might casually refer to as 'modern-looking', probably owes something to the Bauhaus designers.

The Bauhaus was established in 1919 by the state of Weimar

in Germany, perhaps in a spirit of rebuilding after the Great War. Its director was Walter Gropius, and he turned initially for a design philosophy to the Arts and Crafts Movement and the ideas of William Morris. In his 'Proclamation of the Weimar Bauhaus' he declared:

> Let us create a new guild of craftsmen, without the class distinction which raises an arrogant barrier between crafts-man and artist. Together let us conceive and create the new building of the future, which will embrace architecture and sculpture and painting in one unity and which will rise one day toward heaven from the hands of a million workers like the crystal symbol of a new faith.

Twentieth-century architects have been much given to such quasi-religious exhortations and proclamations, based in the still unproven conviction that good design can change societies for the better. In the case of the Bauhaus the conviction was that suitable designs for industrial products would improve the world.

This initial emphasis on craftsmanship soon proved to be unrealistic in what was already a machine age of engineers, and in a proclamation in 1923 Gropius reformulated the aims of the Bauhaus declaring that 'The Bauhaus believes the machine to be our modern medium of design and seeks to come to terms with it'. This was more than mere functionalism, for the responsibility of the Bauhaus was 'to educate men and women to understand the world in which they live and to invent and create forms sym-bolising that world'. In other words they were to discover proto-typical designs suitable for machines and for mass-produced goods. And this is exactly what they did. Following the lead pro-vided by the Deutscher Werkbund, by the Fagus Factory and the Futurists, by unselfconscious industrial buildings such as grain elevators, and by the minimal patterns of contemporary abstract art, they used geometric shapes and plain surfaces as the essential features of the new appearance for machine things. The simpler the lines and forms were, the better they were held to symbolise the modern machine world. This was the style of modernism. There is little in the modern urban landscape that does not bear its mark.

In 1926 the Bauhaus school moved to Dessau and into a new building designed by Gropius and itself called the Bauhaus

(Figure 6.4). The materials were glass, steel and concrete, the forms angular, the appearance funtional and modernist, the organisation and layout cubist. Like a cubist sculpture it had no clear front, but was meant to be seen from multiple perspectives as one moved around it or even through it on the service road that separated the two main wings. With workshops, studios, classrooms and a small apartment block it served many purposes, though its plain surfaces and forms did not clearly reveal what these were. This building was the realisation of architectural ideas that Gropius and his colleagues had been formulating for some years. It was the first mature synthesis of all the architectural movements against useless ornament and for machine-age design, for futurism, for functionalism. It set the standard and it determined the main course of progressive architectural design for the next 50 years.

Figure 6.4: The Bauhaus in Dessau, by Walter Gropius, 1925, with workshops on the left and apartments on the right. This is probably the single most important building and institution in the development of modernist architecture because it synthesised all innovative developments of the previous two decades into a style which has dominated institutional and corporate architecture ever since

Source: Hitchcock, Fig. 161a, original in Museum of Modern Art

In 1928 Walter Gropius resigned as director and was replaced by Hannes Meyer. Meyer's main concern was with developing socially responsible designs and methods of production; under his direction considerable emphasis was placed on inventing satisfactory types of housing for workers. This socialist inclination did not sit well in the increasingly repressive right-wing political climate that was developing in Germany, and in 1932, after a short period under the directorship of Mies van der Rohe, the Bauhaus was closed.

Modernist housing projects

In the years following World War I there was a serious housing shortage, and with the election of socialist municipal administrations in a number of European cities, especially in Holland and Germany, new forms of buildings for public housing were developed. These usually consisted of groups of apartment blocks, rarely of more than four storeys, with laundries, playing fields and other facilities incorporated into them. Many of the members of the Bauhaus were sympathetic to socialism, and were interested in these projects both for their design and their social benefits. After about 1925 they began to turn some of their attention to matters of urban planning and housing. These were systematically and rationally examined in the way that the Bauhaus used for all its design projects. Minimum standards were established for kitchens and bathrooms, the efficiency of the interior circulation of apartments was considered, careful studies were made of the relationship of building height to sunlight and open space, and the possibilities of mass-producing dwellings and of using standardised parts were examined. The built results of these studies usually were ranks of uniform slabs of almost identical walk-up apartments, each one equipped with the sorts of domestic facilities which we have come to expect, such as fitted bathrooms, but which were then still considered remarkable (Figure 6.5). At Frankfurt-am-Main between 1925 and 1931 the city administration encouraged the city architect, Ernst May, in the construction of 15,000 low slab apartments for workers, each apartment having its own efficiency kitchen. This was the inception of modernist town planning.

In 1927 a housing exhibition, the Weissenhof Seidlung, was held at Stuttgart, and most of the leading modernist architects of Europe designed model buildings for it. These were all angular and unornamented. They were also light, airy, functional, and a serious attempt to find ways of improving urban living conditions, especially of the working classes. The hope was that any one of the buildings could serve as a prototype for mass production, for it was believed by the modernists that houses should, as Gropius

Figure 6.5: Modernist planning. Diagrams by Gropius illustrating that widely spaced slabs of eight to twelve storeys leave more open space than two-storey buildings while accommodating the same number of people. A demonstration housing project for workers, by Mies van der Rohe, 1928. These rationalist planning theories and experiments had an international impact on the layout and design of apartments and especially on public housing

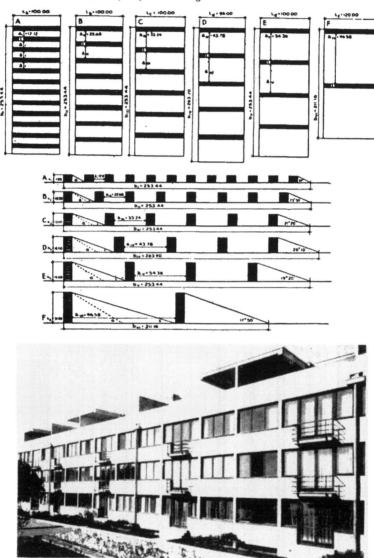

Sources: Gropius, 1965, p. 104, and Hitchcock, 1958, Fig. 162b, original in Museum of Modern Art

put it, be as easily mass produced as shoes; it was simply a matter of finding the right techniques for doing this and then the persistent housing shortages could be solved. This attitude can be criticised for its paternalism — here were affluent, well-educated architect/socialists inventing good housing units for the poor and poorly educated workers, and at the same time imposing a modernist aesthetic upon them. But at the time the motives of those involved were probably genuine enough because they had an unshakeable belief in the merits of modernism and its potential for solving deep social problems. Anyway, regardless of motives, this housing exhibition had a far-reaching effect. The architectural historian Leonardo Benevolo has written of it (1982, p. 486) that this 'was not a collection of proposals for buildings, but a suggestion for a new concept of living, which set out to modify not only single dwellings but the whole of the urban scene'. And to a very great degree it succeeded, if not with the whole urban scene, at least with the landscapes of public and state housing.

The Stuttgart exhibition and contemporary developments, such as Walter Gropius's theoretical studies which showed that apartment blocks of eight to twelve storeys were preferable because they leave the largest usable areas of open space at ground level, comprise the origin of what has been called 'the open block'. This is an arrangement of apartment buildings standing on their own open site, rather than aligned along existing roads. The open block allows the buildings to be arranged in any configuration the planner chooses, and in a world imbued with modernism that invariably meant in rectangularly ordered and well-spaced rows, probably with a mixture of low slabs and slightly taller towers. The open spaces are devoted to parking, to parks and playgrounds.

With the economic depression and with the rise to power in Germany of the National Socialists, who preferred romantic vernacular revivals to modernism, these sorts of modernist projects were built rather less widely than many of their proponents must have hoped. In the early 1930s some open block landscapes were created in housing developments on the fringe of Amsterdam; after World War II they became one of the commonplace features of cities, used especially for public housing. There have of course been modifications to the style and scale of the apartment buildings, and refinements to the layouts, but the essential forms have remained the same — standardised units in look-alike blocks and slabs arranged in orderly ranks (Figure 8.2). They can often be seen in the background to television news reports or photographs from Warsaw, Toronto, Tokyo, Rome, Moscow, Stockholm, London and countless other cities.

Le Corbusier

Paralleling the development of the Bauhaus, articulating equivalent arguments about machine age design and conceiving buildings no less modernist, was Le Corbusier (his actual name was Charles-Edouard Jeanneret, but he took the pen-name Le Corbusier and it is by this name that he is universally known). Along with Frank Lloyd Wright and Mies van der Rohe he is generally considered to be one of the great architects of the twentieth century. Born and educated in Switzerland, he worked briefly in the office of Peter Behrens where he encountered the ideas of the Deutcher Werkbund about mass production and standardisation which were to infuse much of his subsequent thinking. He was deeply involved both in architecture and in urban design and planning. His buildings and his polemical writings were nothing if not rousing. That he influenced the character of modern landscapes is indisputable, though exactly what he influenced and how is difficult to say. His elegant sculptural designs and grand plans have been reproduced, but usually in combination with elements of the relatively simple geometries of the Bauhaus so that it is rarely possible to attribute exact influences for the design of a particular building or urban plan.

The principles that dominated Le Corbusier's early work were not unlike those which inspired Gropius, and he stated them in the modern architect's best manifesto manner in *Vers une architecture* published in 1923. The new architecture was, he exclaimed (in manifestos one can only exclaim or declare), for a machine age, and its elements could already be recognised in industrial products. The engineer's aesthetic, unhampered by styles and customs in its search for efficient design solutions, was the pre-eminent one. Le Corbusier accordingly took as his examples ocean-going liners, aircraft and automobiles, all engineered to serve specific purposes. His architectural inspirations were the buildings of ancient Greece and the grain elevators of modern America, and he included nine photos of grain elevators as 'the magnificent first-fruits of the new age' (p. 33). However, he stressed that engineering alone is not enough to make ARCHI-TECTURE — this requires perfection of profile and contour which are 'pure creations of the mind'. These creations combined with engineering should serve to create 'mass-production houses'.

> If we eliminate from our hearts and minds all dead con-
> cepts in regard to the houses, and look at the question from
> a critical and objective point of view, we shall arrive at the
> 'House-Machine', the mass-production house, healthy (and
> morally so too) and beautiful in the same way that the
> working tools and instruments which accompany our exist-
> ence are beautiful (p. 210).

It is difficult to read this sort of sweeping claim now, in an age attuned to the false rhetoric of propaganda, without wincing at its arrogance and the intimations it contains of the architect-dictator, guaranteeing economic efficiency and moral health for all with one style of mass-produced building. But in the 1920s, after the destruction of the Great War and before the rise of National Socialism and Stalinism, it seemed heroic. Le Corbusier had, in fact, already worked out designs for producing reinforced concrete houses which would take just three days to assemble using mass-produced fittings. This system was never put into complete practice, though he did complete a development of workers' houses at Pessac near Bordeaux in 1925 in a mass-produced, square-set, plain façade style.

In that same year he listed the five points he considered to be essential for a new architecture: (1) Buildings should be raised on stilts, or pilotis, to allow free access beneath. (2) Buildings should have roof gardens and terraces because reinforced concrete and central heating systems made the gable roof obsolete. (3) The floor plan should be open because load-bearing walls were no longer needed. (4) Horizontal strip or ribbon windows should run the length of the building. (5) Façades should be treated sculpturally; this was made possible by the use of reinforced concrete which was the modern material Le Corbusier found most suited his designs. These five elements were incorporated by Le Corbusier into most of his buildings of the 1920s and 1930s (Figure 6.6).

It is impossible to maintain that he was the major source for

Figure 6.6: Villa Savoye, Le Corbusier, 1929. This building illustrates several of Le Corbusier's five points of architecture — the pilotis or columns to allow free access at ground level, ribbon windows, a sculptural façade, a roof garden and an open floor plan

such things as flat roofs, but his designs were undoubtedly an original version of modernism that has been duplicated, usually in somewhat less sophisticated forms, by other architects. Though all of his five features of a new architecture are rarely all found together in a single building, their influence can often be seen in such things as the ribbon windows and sculptured concrete façades of institutional buildings and suburban offices. For houses, however, they have proved unpopular. Not only have relatively few Le Corbusier-style machine-houses been built but even his prototype development of modernist workers' houses at Pessac has undergone many changes and modifications. The free façades have been altered, awnings have been added, porches put over the doors, and windows have been blocked in. As a style for detached houses modernism, whether Le Corbusier's or anybody else's, has not received popular acclaim, and the machine-house has never been mass-produced.

Principles of modernism in the work of Mies van der Rohe

The philosophy which most affected modernism during the 1920s and in its early stages of development was the so-called Neue Sachlichkeit — the New Objectivity. This was a philosophy of detachment and rationality; for architects it implied not just clean lines and geometric order but a systematic working out of design problems that resulted from the new machine technologies and social needs. The major principles of design that followed from this were these:

1. Buildings should be treated as space-enclosing volumes rather than masses; this had been made possible because the new structural materials of steel and concrete meant that walls no longer had to be load bearing and the internal spaces could be divided up according to need and purpose rather than structural necessity. It meant that buildings could be reduced to more or less elegant boxes.

2. The appearance of buildings could be derived from the shapes of their vertical and horizontal elements and their repetition.

3. Technical perfection and fineness of proportion were to be stressed, partly to demonstrate the engineered character of a design and partly to provide aesthetic quality in the absence of all decoration.

4. Most buildings and their environments should have an industrial, technical, mass-produced character, reflecting the ideal of designing for a machine age. This latter idea is often described as functionalism. In fact, it often did not and does not meet the purposes of particular buildings very well; the buildings leak, are noisy, the spaces are awkward. From the point of view of the user modernism most usually seems to be an elitist, *avant-garde* aesthetic fashion thrust upon an unsuspecting and largely uncomprehending public.

The design philosophy of the New Objectivity was nowhere better demonstrated than in the buildings of Ludwig Mies van der Rohe, an architect who, like Gropius and Le Corbusier, had worked with Behrens before World War I and who had joined the Bauhaus in the early 1920s. Possibly more than any of his contemporary architects he aimed for technical perfection and exact proportions in his buildings. His designs reduced buildings to the simplest elements of horizontal line, vertical line and surface. Perfect reason and precise order, as in minimalist paintings of rectangles and lines, were to dominate. In this manner he designed several buildings, but none was more perfect than the Barcelona Pavilion for an International Fair held there in 1929 (Figure 6.7). This was, in all but name, a temple to a new aesthetic, a testament to precise abstract and rational thought. It was a single-storey structure of utmost geometric simplicity, with smooth surfaces of marble and glass; it contained four chairs and a few stools, also designed by Mies van der Rohe, and two classical statues (presumably to suggest classical perfection). Here then was the quintessential expression of 'functional', modernist, rational, machine-age design. It was in fact so rational that, apart from aesthetic contemplation, it scarcely tolerated human use at all because humans tend not to follow straight lines and to find hard, flat surfaces less than comfortable.

Such spare geometries are not easy to copy well, especially in buildings that actually have to be used, but they are remarkably easy to mimic in some less than perfect manner. They have, accordingly, been a primary source for the design of everything from operating theatres and bus shelters to the skyscraper offices of the 1960s and 1970s. The Barcelona Pavilion was demolished soon after the exhibition closed, but its ghost is with us everywhere.

Figure 6.7: The Barcelona Pavilion, Mies van der Rohe, 1929; in effect, a temple to the new modernist aesthetic, its pure surfaces and spaces enclosed only a few chairs and sculptures. In the 1980s it has been reconstructed on its original site

Source: Hitchcock, Fig. 165a, original F. Stoedtner

The international movement

The grain elevators of North America played a prominent role in the development of the thinking of Le Corbusier and Walter Gropius, and Frank Lloyd Wright's early designs greatly impressed European architects, but the fact is that modernism in the 1920s was almost exclusively a European matter. Indeed at the International Exposition of Modern Industrial and Decorative Arts held in Paris in 1925 (and from which Art Deco takes its name) the United States declined an invitation to set up a pavilion on the simple grounds that there was no modern art in America (*Architectural Record*, 1925, p. 373). This sort of parochialism was actually quite out of place in a world in which artists, planners, businessmen and architects had for the previous 50 years been moving freely and frequently from continent to continent, self-consciously exchanging ideas as they went.

From its earliest murmurings modernism had implicitly disavowed all national and regional styles for it emphasised synthetic materials, standardisation and mass production, all of which are uniform and universal. In 1925 Gropius published a short book called *Internationale architektur*. Then in 1932 two American architects, Philip Johnson and Henry-Russell Hitchcock, organised an exhibition of modern architecture for the Museum of Modern Art (established since the Art Deco exhibition of 1925, no doubt in an attempt to catch up) in New York City. Following the lead of Gropius they called the exhibition 'The International Style', and they included illustrations of the work of Mies van der Rohe, Le Corbusier, Walter Gropius and all their European contemporaries, as well as pictures of two modernist skyscrapers and a gas station from North America. It was an exhibition in lavish praise of the new architecture and its rational qualities. Here, it seemed, was a style that in its depth and versatility compared with the best historical traditions, was entirely original, and which was equally appropriate in any climate, culture or city.

The name — The International Style — has not stuck very well. Writing in 1966, in a preface to a new edition of the exhibition guide book, Hitchcock argues that the style had already died; he would presumably reserve the name exclusively for the formative period of modernism and the pure styles of the masters of the 1920s. Be that as it may, there is no doubt at all that modernism in all its derivative forms and copies is thoroughly

117

international. Actually it turned out to be more immediately international than Johnson and Hitchcock could have anticipated, because at the dissolution of the Bauhaus in 1933 its members, depending on their political proclivities, went either to England and America or to Russia. Both Walter Gropius and Mies van der Rohe ended up in America, where, after World War II, their design influence would be almost unquestioningly accepted by the executives of great business corporations and institutions anxious to install themselves in buildings which symbolised progress and prosperity. International modernism was to become the overwhelmingly dominant architectural style in the urban landscapes of the 1960s and 1970s.

7

Landscapes in an Age of Illusions: 1930 to the Present

Modernism and the International Style were obsolescent almost as soon as they were invented. They had been conceived as a style appropriate for the machine age technologies of electricity and mass production at just about the time that an entirely new set of technologies was beginning to be developed, including polymer chemistry (the basis for nylons and pesticides) and television. These were to be followed in the 1940s and 1950s by nuclear engineering, cybernetics, jet engines and electronics, and then by microchips, satellites and genetic engineering. Most of these recent technological transformations have been swiftly commercialised, and they have, Reyner Banham suggested as long ago as 1960, 'visibly and audibly revolutionised' the small things of life. Even then he felt he could justifiably describe this as the Second Machine Age, one based in domestic electronics and synthetic chemistry. There are now synthetic fabrics, plastic bowls, televisions, video recorders, microwave ovens and transistor radios in almost every household, polystyrene coffee cups are thrown away without a second thought, and cash comes from computers in the street.

The technological innovations of the late nineteenth century had a swift and obvious impact on the look of cities; structural steel made skyscrapers possible, electricity changed the appearance of night, automobiles led to all the paraphernalia of the highway landscape and to the low-density residential suburb. Since many of the Second Machine Age technologies have now been in use for 40 years it is surely legitimate to consider what their landscape effects have been. Their social consequences certainly seem to be considerable. For instance, Joshua Meyrowitz (1985, p. 76) reports that in the United States people watch television for an average of 30 hours a week. This amounts to about a third of one's waking life, and regardless of concerns about content such devotion to a single, new and passive activity has to have considerable social importance. Meyrowitz argues that electronic media, especially television, have indeed changed the 'situational geography of social life' (p. 6). Telecommunications flow indiscriminately through all regions, they put hitherto isolated places into direct contact with the electronic centres of influence, and they 'bring information and experience to everyplace from everyplace' (p. 118).

The deep effects of these sorts of social changes are summarised in phrases like 'the information society' or Daniel Bell's term 'post-industrial society'. Such expressions are presumably meant to imply that an original type of society and economy is emerging, one based less on neighbourly association and the manufacture of goods than on the electronic processing of information, the provision of services and globally shared electronic experiences. Equivalent social and economic changes are also often held to have followed from computers, and to a lesser extent from mass international tourism and the widespread use of synthetic chemicals. Surely by now there should be clear and extensive manifestations of these changes in urban landscapes. Yet we look for them almost in vain. By far the most obvious visual fact about the parts of cities made since 1945 is that they are extensions and expansions of types of landscapes which were developed in the previous half-century. Very little in them seems to be wholly new, just taller and sleeker skyscrapers, more suburbs and more expressways. It has to be concluded that what seems to be happening is that, unlike the older technologies which quickly led to major visible changes in cities, these new technologies neither require nor suggest particular built-forms and landscape features. Instead their special accomplishment is to turn places into huge equivalents of plastic flowers: things are left looking much as they always did but their materials and meanings are profoundly changed. From the perspective of landscapes it seems that this is not so much a Second Machine Age as an age of illusions.

Microscopic manipulation and anonymous expertise

The technologies of the Second Machine Age seem to leave no

clearly identifiable footprint of their own, and this is perhaps because they have to do with information, substitution and processes occurring at microscopic scales. The processes of polymer chemistry, nuclear physics, genetic engineering and electronics are invisible to the naked eye and mostly incomprehensible to the non-scientific expert. Radiation, chemical bondings, electrons, and the use of enzymes to slice and splice DNA chains are matters which we have to accept on faith and the evidence of the products we believe to have been created by them.

One consequence of this microscopic manipulation of natural processes and materials that is directly relevant to landscapes is that the new technologies tend to insinuate and adapt themselves to existing forms rather than to generate new ones. Now, it is often the case that new technologies mimic the forms of whatever preceded them, hence the first cars were made to look like horse-drawn carriages. But this time it may be different because the essential fact of these microscopic technologies is their malleability (Figure 7.1). In other words, the fact that there are so few architectural and landscape forms which are a product of these technologies cannot be dismissed as merely a delay in discovery.

Figure 7.1: Plastic magic. An advertisement for a VISA card in Melbourne Airport in 1985. The technologies of the Second Machine Age seem to have an ability to mimic any form while suggesting few distinctive forms of their own

Because the technological processes are invisible their constructive effects will also be as invisible as harmful nuclear radiation or toxic chemicals in rivers. The structure and quality of things is profoundly changed, their appearance remains almost untouched.

A second consequence of microscopic invisibility is that much of what happens in the technologies of the Second Machine Age is beyond the grasp of the layman. Even the systematic workshop methods of trial and error invention used by Edison and his contemporaries have little role to play in any of this. Technological discoveries now are the result of precisely orchestrated programmes of scientific research aimed at achieving specified practicable outcomes. There are no inventor heroes in the Second Machine Age, and no day trips out of New York City to look at the modern equivalent of light bulbs. Millions of us may benefit or suffer because of some new electronic or genetic process which only a handful of mostly anonymous scientists and technical experts can really understand. So popular appreciation, if it occurs at all, is likely to be restricted to a few moments reading some ecstatic account of the new marvel in a glossy magazine. There can be little question that for most people modern scientific technology is little short of magical.

Visible effects of new technologies

Actually it is overstating the case to claim that these microscopic technologies have had no visible consequences. There are a number of specific landscape features which can be attributed to them, though these are striking chiefly because they are so much more limited and localised than one might expect. The most apparent of them are the following.

Plastic backlit signs

With the development both of plastics and acrylic paints the flickering glow of inter-war neon and argon signs was replaced by the steady shine and primary colours of backlit signs (though neon has recently been revived for window and interior decorations). In dramatic locations, such as Times Square, individual signs can be up to a block long, but in any commercial area, suburban strip or main street, backlit signs are now the major sources of bright colour.

The paraphernalia of telecommunications

These include TV antennas, transmission towers, satellite receiver dishes and microwave towers. Their forms have changed as the technology of telecommunications has changed. First there were fields of radio transmission towers, then metal-frame tower television transmitters, and the spiky forests of TV aerials on houses which so deeply offended some aesthetic sensibilities. These have recently begun to be replaced by underground cables and satellite dishes. Microwave towers for telephone and television signal linkages have sprouted from the roofs of skyscrapers and from hilltops in a variety of metal frame and concrete tower forms since the mid-1960s. I suspect that every commanding hilltop in the populated strip of Canada has one. No city with any claim to being progressive can be without a telecommunications tower — the GPO tower in London, Centrepoint in Sydney, the CN tower in Toronto. These are the most prominent landscape objects of the Second Machine Age. They are also obsolescent, having been bypassed by the much less visible technologies of satellite communications.

Computer manifestations

I know of only two ways in which computers are directly manifest in urban landscapes. The insta-bank, money machine, or electronic teller is a miniature computer that reveals the dependence of banks on electronic account processing. These are significant not only because they have become such a prominent feature of the streetscape with lines of people waiting to indulge in electronic intercourse, but also because they have taken what was previously a very discreet and personal service of almost confessional quality, and turned it into something entirely public and impersonal. There can be no better indication of the role of money in modern culture.

The second indication of the effect of computers is more subtle. There are certain concrete structures, notably those with complex curves such as domed stadiums, which probably would not have been constructed were it not for computers because the calculations of stresses and bar-bending schedules, though technically possible, would have been too time consuming and costly. In this case, as with energy-efficient structures in which computer systems control heating and ventilation, computers have made modifications to built-forms possible rather than inspiring entirely original structures.

Airports

Airports were, of course, first built in the 1920s and 1930s, but they have so greatly increased in scale and complexity that they are really post-1945 landscape features. They are dependent on a number of new technologies: radar, computers to monitor flight traffic and to arrange seating, fire-retardant foams, and the jet engines which make air travel fast and attractive to businessmen and tourists. Airports are one of those modern built-features for which there are no clear design precedents, and in that sense they are as much symbols of the twentieth century as skyscrapers. However, since their primary landscape characteristic is the open space devoted to runways, and since their buildings are treated by designers and passengers alike as places of transit to be departed as quickly as possible, they seem not to attract much attention as visual entities. For all the sophisticated new technologies that contribute to their existence airports are, I suspect, merely bigger, noisier, more crowded equivalents of their 1920s predecessors.

Nuclear energy

Nuclear power stations and the research facilities for nuclear physics have created new landscapes, though in the latter case the linear accelerators and cyclotrons are so few and so out of the way that they are only likely to be seen if one makes a deliberate effort to visit them. Nuclear power stations, too, are often hidden away, probably for security and political reasons. All one can see of the Pilgrim plant in Massachusetts of the San Luis plant in California are security gates and a short stretch of road. Those more prominently located have distinctive geometric forms with spheres, domes and large cubes, and more or less elaborate security systems. Beyond the distinctive structure their landscape influence disappears; the electricity they contribute to the national grid, and all the elements of pylons, transformers, poles and wires, are as they would be for any other type of generating station.

Agricultural landscape

Though it is not strictly relevant to a discussion of urban landscapes, it is worth noting that new technologies have had a marked impact on the agricultural landscape. The long narrow sheds of broiler farms, and the great, chemically-green fields devoted to a single crop exist only because of antibiotics, pesticides and fertilisers, and possibly because of computers to monitor the applications of all of these.

Even when taken in combination these features amount to little more than isolated and superficial landscape changes. They certainly do not comprise a new type of landscape for a new technological era in anything like the same way skyscrapers, highways, commercial strips and modernism had by the 1920s begun to create a landscape for the first machine age. It can be argued that this is not because of the microscopic character of new technologies but because so much of the existing fabric of cities has been made by technologies of the first machine age, and it is simply too soon to look for major modifications to accommodate the newer technologies. Peter Blake, for example, maintains that the effect of information processing on cities will, in due course, be profound. He speculates about decentralisation, the decline of office-based work and the associated commuting to city centres because computers can be located anywhere; shopping by closed-circuit television will 'transform the configurations of today's and yesterday's shopping centres and cause grass to grow on the vast highway and parking facilities generated by them' (1982, p. 159). He also expects shifts in intra-urban, suburban and rural patterns, though he does not elaborate about these.

Using a little of the eye of faith it is possible to recognise early signs of some of these shifts. 'Lights-out' buildings are fully computerised factories in which electronic components are produced in virtual darkness; still rare they are likely to become increasingly commonplace. 'Back-offices' are branches of corporations where some of the more menial data-processing tasks, such as dealing with payments for charge accounts, are carried out, usually by women working in great hangar-like buildings, sleek on the outside and filled with ranks of computers on the inside. These are often located in the suburbs where they are more accessible to employees, and connected electronically to other branch offices and to the head office which remains downtown. Such back-offices indicate that some decentralisation is occurring, as do the mini-downtowns of offices and shops which have recently been developed in suburban districts of metropolitan areas such as Toronto and Houston. It is, however, difficult to maintain that electronic automation and communication are more than a minor cause of such decentralisation, operating in association with shifts in land and labour costs, expressway construction and, above all, accessibility to the new sites by automobile.

A much more obvious and important landscape fact is that the styles of the back-offices, and indeed of the front-offices, computer factories, data-processing centres, TV studios, bio-engineering laboratories and missile component plants, give little or no indication of the activities being conducted within (Figure 7.2). For most of these activities almost any convenient building, old or new, in almost any reasonably accessible location will do. The new ones tend to be stylishly modernist, neatly landscaped, with a corporate sign or logo outside that reveals nothing to the uninitiated — Apple, Nielsen, Litton; the old ones may be neatly renovated, though there are no rules about that. It seems that for the most part the technologies of the Second Machine Age neither require nor suggest specific landscapes.

Adaptability, mimicry, opacity, ephemerality

The primary impact of the new technologies at the scale of landscapes, in so far as it can be recognised at all, has probably been to extend the possibilities of older technologies by making them cheaper and more efficient. They have significantly reduced constraints of cost and application. In achieving this result they have had remarkably few direct effects on the appearance of buildings and on urban layouts. Just as plastics and fibreglass can be used to make copies of oak beams in reproductions of English pubs, fake brick fireplaces, and even fake steel beams, but suggest few

Figure 7.2: Opaque architecture. An Apple Computers building near Palo Alto in California, and the Litton Building in Toronto where guidance systems for Cruise missiles are manufactured; electronic technologies disappearing into modernist buildings which give no indication of what goes on within

distinctive forms of their own, so computers, bio-genetics and nuclear-generated electricity adapt to or reproduce whatever structures and landscape arrangements already exist. As a consequence the changes they induce in landscapes often provide few clues about causes. The manicured lawn of a suburban house could be the organic result of diligent watering and hand weeding, or it could be achieved by having a company called Chemlawn come in once a month to spray herbicides and fertilisers; an ancient-looking medieval cathedral can be kept in that condition by a virtually invisible preservative sprayed on to the stonework; a fibreglass replica of Stonehenge was seriously proposed (and the idea subsequently dropped) in order to protect the original from the wear and tear of tourists; there are concrete fences that look like wood, plastic trees and plastic thatch, artificial turf, all virtually indistinguishable from the real thing, at least at a distance.

On a larger scale the microwaves, fibre optics, information flows, automated offices and data bases disappear into existing skyscrapers, suburban offices, renovated victorian mansions and remote cottages, drastically changing their interiors but leaving their exteriors untouched. And as tourists, businessmen, electronic messages, credit cards, corporations and synthetic chemicals circle the globe they generate international airports, expressways, skyscrapers, convention centres, billboards and resorts which are remarkable only for their global sameness. The forms of these globally undifferentiated landscape features almost always derive from older traditions and technologies; it is only their diffusion and geographical confusion which are new.

So the products of the new technologies are adaptable, they can mimic other materials and processes, and they can hide. They fit well into the opaque modern city where most of the serious problems are out of sight, poverty is hidden in trim modernist apartment blocks, pollution is almost always invisible. They also fit well into the restless ephemerality of modern society. 'Development' and 'redevelopment' must be two of the most commonly used words in association with cities; urban places are not permitted to stay as they are for more than a few years, they must be developed, redeveloped, renovated. Perfectly sound office buildings constructed less than 30 years ago have to be replaced by newer offices, employees have to retrain, workplaces to be retooled, everything brought up to date, façades remodelled in the latest style. Cycles of fashion in decoration and

128

in architecture seem to be becoming shorter and shorter. A sense of urgency and speed runs through much of modern urban life. Each generation expects not to hand on a tradition but to be rendered obsolete, along with its artefacts and environments, by the next generation.

How difficult it is to make any sense of all of this. The old looks of building now can (a) disguise brand new scientific laboratories within, (b) be reproduced in any one of several new materials, (c) be renovated to and preserved in a state of pristine oldness, (d) indicate that a speculator is holding the site for future redevelopment, (e) be changed to some other new old appearance. Neither traditional nor modernist standards offer much help for understanding or criticising what is happening. It seems all we can do for the moment is to remain alert to the variety of possible deceptions and the problems they hide while we try to grasp the character of the process and the society which produces them.

Imagineering

'Imagineering' is Walt Disney's word — a contraction of imagination and engineering. It neatly captures the character of the scientifically and technically based creativity that stands behind the illusions of the Second Machine Age. Imagineering is most concentrated and most obvious in places like Disneyland, in television productions and on movie sets. The production of even the simplest TV programme requires a huge electronic backup of cameras, control booths, transmission and receiving devices, almost none of which is apparent on the moving images on the screen in one's living room. At Universal Studios in Los Angeles entire outdoor sets are 'winterised' in the middle of summer, great black backdrops allow day to be turned into night, and there is a computer designed park set with real grass and little metal stumps and holes to which fake trees of any species desired can be attached in ways that permit over 200 camera angles.

The illusions of modern landscapes reach far beyond the television studio and movie set. There are, for example, the illusions of history in carefully constructed museum-villages like Plimouth Habitation, a mock-up of the Pilgrim Fathers settlement in Massachusetts. Painstakingly researched to be accurate in every detail, including the seventeenth-century dialect spoken by the apparent residents, and surely of educational value, this is inevitably idealised by omitting dirt and disease, and is in fact not on the original site which is occupied by the real town of Plymouth. Considerably less technical, but scarcely less illusory, are sales pitches for new suburbs which reek with images of gentility and romanticism, selling the life-style of TV soap operas. These images are made real in such things as great gateway entrances to housing developments, brass coach lamps and door handles, double-door front entrances, and interlocking brick driveways.

It is but a short step from illusion to confusion. An advertising sign identifies a new development in Toronto as having 'Historic Homes of the Future'. The centre of the village of Gibsons, on the coast of British Columbia, is dominated by a restaurant called Molly's Reach which is in fact the set for the television series 'The Beachcombers' and is not a restaurant at all. The mining town of Kimberley, also in British Columbia, having fallen on hard economic times, decided to 'bavarianise' itself in order to encourage tourism; the city fathers provided funds for suitable façade changes, a street was closed to make a Stadtplatz and the high school band performs there wearing lederhosen. And in Hollywood the Sidewalk of the Stars terminates in a grand star honouring the first Apollo mission and Moon landing (Figure 7.3). Illusion, fiction, reality, science and technology have begun to merge indistinguishably in landscapes.

So it continues; illusions like these can be found in almost every arena of modern life and, in fragments at least, in almost every environment. Of course all cultures have myths which serve to explain existence and give meaning to life. The difference between these and modern illusions is that the latter are deliberately manufactured and manipulated, often solely for economic and political ends. Their basis lies not in shared social experiences, but in technical knowledge. Imagineering, the imaginative engineering of deception, has become one of the primary ways of making landscapes.

The megamachine with its electronic eyes on the street

Behind the mimicry and the imagineered landscapes of the Second Machine Age there is a complex organisation made up mostly of engineers, economists and accountants, executives and

Figure 7.3: Imagineering the landscape. Molly's Reach in Gibsons, British Columbia, is not a real café but a set for the TV series 'The Beachcombers'; a brochure for the 'bavarianised' town of Kimberley in the Canadian Rockies; 'Historic Homes of the Future' are (will be?) in Toronto; the Apollo Mission Star in the Sidewalk of the Stars in Hollywood. Fiction, reality, science and illusion become inextricably merged

politicians. These are the manipulators of data and the inveterate supporters of progress through technology, those who welcome the post-industrial society. Artists, poets, novelists and philosophers have not often shared this enthusiasm. In the 1890s it may have been possible for Bellamy and Morris to dream of a utopian future based on electrical technologies and socialism, but for most of the twentieth century artists have expressed despair at its very thought. The image of the technological future offered by E.M. Forster in 'When the Machine Stops', Eugene Zamiatin in *We*, Fritz Lang in his movie *Metropolis*, Aldous Huxley and George Orwell, is thoroughly depressing: repressive governments, violent societies, ugly places, loss of personal identity and of any purpose in life, individuals drifting through pointless routines yet kept content because they are sated with material and sensual pleasures.

Lewis Mumford shares the artists' concerns. The drab and repressive world which they anticipate is for him the inevitable outcome of what he calls the 'Modern Megamachine'. His argument is this. All urban civilisations have been based.on the development of a machine-like system of administration, for this was necessary to orchestrate large-scale projects of construction and distribution. The modern megamachine is an especially sophisticated version of this system. It has been developing for about 300 years, but only with the principles of organisation developed during the scientific projects of World War II, and with the subsequent availability of electronic devices for the storage and communication of data, has it attained its complete realisation. These have enabled all modes of power — administrative, legal and physical — to be reduced to standardised procedures. In particular, computers, with their instantly accessible and unforgetting memories, allow the integration of many administrative sub-systems into a single large system in which the benefits of all the individual sub-systems (such as greater output and accessibility to information) are increased. But so, too, are all the disadvantages, including rigidity, lack of response to new situations, detachment from human purposes, and control of local issues by remote centres of authority.

The modern megamachine is an enormous, unthinking device for ordering, organising, and controlling everything it can. These efforts, Mumford argues, have no human purpose. The Pentagon of Power is the metaphor he proposes to describe this — for, like the Pentagon, the megamachine is insensitive to information, especially qualitative information, which does not fit into its system, and it appropriates to itself expertise and authority and money for purposes which fall beyond the bounds of human reason. It has only one efficient speed — faster; only one attractive destination — further away; only one desirable size — bigger; only one rational goal — more.

If Mumford is right the products and manifestations of the megamachine include skyscrapers, atomic reactors, suburban tract developments and shopping malls, new towns, underground control centres, international airports, theme parks and space shuttle launching pads — in short, almost everything in modern landscapes. It should be obvious from the scale and complexity of all of these that they depend on highly sophisticated technical knowledge and organisation. Yet somehow awareness of this fact seems to be in the background, possibly because we are distracted by imagineered illusions or perhaps because we are usually more concerned with landscapes themselves than how they got there. Nevertheless, it is occasionally possible to catch small glimpses of the megamachine almost directly at work. For instance, the militaristic obsession with security and order has apparently been transferred to business corporations, and these are now taking sophisticated security measures to protect themselves against industrial espionage, terrorists, vandals, distraught customers, or any other socially or politically unreliable individuals. In commercial and industrial areas of the modern city businesses enlist security forces, establish control booths, sometimes remove all logos and names from their buildings, and then set up remote video cameras and electronic alarm systems (Figure 7.4). Closed-circuit video security systems have become a matter of course for security-conscious corporations and government departments. On rooftops in Westminster, on neo-gothic pinnacles in Ottawa and neo-classical cornice lines in Washington, overlooking entrances and silently scanning the interior of shopping malls, the remote control video camera is a symbol of security. It is a reassuring indication that here at least the megamachine is going about its business of maintaining control. It is the megamachine's electronic eye on the street.

Nuclearism

Lewis Mumford argues that a major consequence of World War

Figure 7.4: Electronic eyes on the street. Video security at a data processing centre for the Bank of Nova Scotia in suburban Toronto. The building is entirely anonymous, presumably to confuse would-be thieves, and has all its approaches continuously surveyed by video cameras, of which just three are visible in this photograph

II was the development of a military-industrial scientific establishment, and he sees this as being essential to the growth of the megamachine. He also suggests that 'in order to keep the megamachine in effective operation once the immediate military emergency was over a permanent state of war became the condition for its survival' (1970, p. 369). This cold war has centred on the accumulation of nuclear arsenals and increasingly sophisticated delivery systems. All the problems and hopes of modern society are conditioned by the possibility of nuclear war. This has created the greatest landscape illusion of all. Modern urban landscapes might look as stable and enduring as their predecessors but they are in fact continually threatened by the possibility of complete annihilation. For all their enormity and solidity, for all their electronic sophistication and imagineered settings, modern cities are the most fragile cities in history.

By serving as deterrents against the destruction which they would themselves cause, nuclear weapons have paradoxically become our saviours. The threat they pose, and the fact that they protect us from this threat, are constant companions to our lives. This is an almost religious phenomenon simultaneously embracing ideas of possible apocalyptic destruction and salvation. Robert J. Lifton calls this 'nuclearism'.

There are obvious ephemeral indications of nuclearism in protest marches, TV reports of nuclear bomb tests, and aircraft practising their under-the-radar flying. Enduring manifestations are the missile silos, defence installations, and indeed all the paraphernalia needed to fight a nuclear war (in a different jargon — to maintain deterrence). In spite of the stupendous quantities of money poured into these and their enormous significance, they are largely hidden from public view. Even when they are visible they are not obviously recognisable; missile silos consist of a few pipes and valves and a concrete stairwell and a security fence — everything of significance is below ground. Nuclearism, like so many other aspects of the technologies of the Second Machine Age, leaves few visible marks in landscapes (Figure 7.5).

Indeed, what is perhaps most striking about nuclearism is that while it fundamentally alters the meaning of existence and conditions virtually everything we think, it seems to change nothing of substance. Life continues as normal. Nuclearism raises deep and persistent doubts about human and personal values and survival, about links with the past, about the viability of nation states, about the accomplishments of science and technology. Yet we do survive, our children go to school, cities and nations persist, the structures of authority reinforce themselves, and scientific technology inserts itself ever further into the interstices of our everyday lives.

In American cities public buildings still have faded little yellow and black signs designating them as fall-out shelters. These are nostalgic reminders of a time in the 1950s when nuclear war was thought to be survivable by such simple means. It was a time when the American Institute of Planners urged research to determine the characteristics which made urban areas 'remunerative' or 'unremunerative targets' for nuclear weapons, and to find ways in which the unremunerative layouts 'can be combined with characteristics that promote efficient industrial production, economical city operation and pleasant urban living'

135

Figure 7.5: Landscapes of nuclearism. The Missile Park outside the missile testing base at Point Mugu near Oxnard in California; and the monument to the first self-sustaining nuclear chain reaction on the University of Chicago campus, sculpture by Henry Moore

(American Institute of Planners, 1953, p. 268). Propaganda efforts are still being made to persuade us that such measures might keep some of us alive, but the real fact is that in a world of surplus nuclear warheads and intercontinental ballistic missiles any city, indeed all cities, can be easily obliterated. This simple fact tarnishes all the mirror-glass towers and spreads an invisible cloud of doubt over the tracts of suburban houses, irreversibly changing their meaning even though their appearance remains untouched.

8

Planning the Segregated City: 1945-75

The connection between cities and planning is now so completely established that it is something of a surprise to recall that before World War II town planning was little more than the exotic concern of a group of idealists, an activity without much legal bite and on the fringes of urban government. Immediately after the war it moved to centre stage, was given a firm statutory basis, and by 1950 had become the chief means for controlling changes in urban development. Before the war a planned development was still a rarity, a model to be aspired to; after the war all developments were planned, at least in some degree. The consequence is that, unlike their antecedents, the urban landscapes made in the late twentieth century are clearly the by-product of planning. This is not to say that their appearance is itself planned; on the contrary planning works mostly in the two dimensions of maps and has little concern for the overall look of things. It does, however, directly affect land-uses and the layouts of buildings at every scale from the smallest items of street furniture up to urban regions, and it therefore conditions the appearance of cities. Its most distinctive marks, those least diluted by existing developments or the claims of private enterprise, are to be seen in residential developments subject to the use of standardised planning procedures, in urban renewal projects and pedestrian precincts, in new towns, and in the forms and patterns of modern highways.

There were several reasons for planning's sudden rise in status in the late 1940s. One was the urgent need for reconstructing war-damaged cities; another was that wartime projects, like the Manhattan project and the organisation of food supply, had demonstrated the merits of large-scale, centralised administration. In Britain, and Britain was a leader and innovator in

138

these matters in the immediate post-war period, the acceptance of planning was also encouraged by the election of a socialist government predisposed to the idea of state involvement in social and economic life. The British New Towns Act of 1946 and Town and Country Planning Act of 1947 provided legal mechanisms for the creation of new towns and for local authorities to control development in their districts, and these were a model for planning practice to which other countries aspired. In Canada, Holland, Sweden, France, Italy and Germany enabling legislation was either revitalised or passed soon after the end of the war to make municipal planning an everyday fact of life. In America planning established itself somewhat more slowly and less obviously into the daily life of cities, coming notably with the Federal Housing Act of 1949 which gave the federal government extended powers over residential development by attaching conditions to the granting of mortgage monies.

There was a noble philosophy behind all of this. Thomas Adams, a widely respected planner who had practised in Britain, Canada and the United States, wrote in 1932 (p. 14) that 'the general object of planning must be to promote human welfare — health, safety and convenience, so far as this can be done by securing order and balance in the physical growth of communities'. Here was a possible motto for post-war planning everywhere. Health, order, safety, convenience, growth, are words that occur again and again in planning documents; the duty of planners was to create a new urban order out of existing chaos and there were no doubts that this could be done. Thomas Sharp's *Oxford replanned*, a report prepared between 1945 and 1948 at the request of Oxford City Council, concluded with the claim that it was 'a plan to shape and direct the future of Oxford ... a plan to preserve old beauty and to make new beauty possible; to add new convenience; to achieve, for the first time in the history of the city, a social balance and a functional equilibrium' (1948, p. 180). Planners in the immediately post-war period were not wanting for confidence.

This confidence lasted well into the 1960s, and then began to wane, especially as serious doubts arose about the social and visual merits of large-scale urban renewal, and as planners discovered that most of their time was taken up in coping with the day-to-day concerns of sewerage systems, tree preservation orders, land use surveys and the administration of development applications. About 1975 — there is no clear turning point and in some cities it was much earlier, in others somewhat later — planning seemed to become less sure of itself, less authoritarian and more sensitive to community needs and the existing urban scene. This change is coincidental with shifts in architectural fashion and social conditions and is better examined in the context of those in a later chapter, rather than as an extension of those years between 1945 and 1975 when planners wielded a new and mighty tool with few doubts about its effectiveness in rebuilding cities.

Standardised planning procedures and planning by numbers

Post-war planning legislation gave to local municipalities the authority to direct the character of change in most aspects of their built environments. There were three general mechanisms for doing this. First, municipalities had a statutory responsibility to prepare an official plan (sometimes called a development plan) consisting of a set of documents and maps which established guidelines for future development and intended land uses; this gave control over large-scale urban patterns. Second, municipalities were given powers to expropriate private property and to redevelop it where this was warranted because of war damage or slum conditions; this permitted planned reconstruction and renewal projects in city centres. And third, municipal planners had a responsibility to make sure that all development proposals conformed to the guidelines set out in the official plan; this meant that all new building, whether in the inner city or in the suburbs, was made subject to sets of standards that covered everything from the layout of neighbourhoods to the design of curbs and the size of windows.

In his book *People and plans* (1968) the sociologist Herbert Gans observes that a typical development plan has sections on major land use zones, transportation and open space, a master plan map, and a 'rhetorical appeal to citizens and politicians to participate in and support the realization of the plan so as to achieve an orderly, efficient and attractive community' (p. 60) (an accurate summary of the official plan for the part of metropolitan Toronto in which I live). Such plans have, he claims, been prepared on an assembly-line basis.

This assembly-line standardisation occurs in all aspects of

town planning. It resulted partly from overarching legislative guidelines, partly from the urgency for reconstruction in the years immediately after the war, and partly from the administrative difficulties involved in planning for rapidly expanding populations. Planners had little choice but to adopt a limited repertoire of ideas and procedures, almost all of them conceived before the war. They included the zoning of land uses, neighbourhood units, Bauhaus-style layouts for public housing, and, in Europe, a new idea — the traffic-free pedestrian precinct. These were widely, even internationally, borrowed, modified slightly, applied, then borrowed back again. There is little in late twentieth-century planning which is not international.

For reasons of convenience and necessity most planning procedures have been reduced to guidelines and listed in manuals provided by professional organisations or agencies, for example there are the *Community builder's handbook* published by the Urban Land Institute in the United States, or De Chiara and Koppelman's *Urban design criteria*, and a multitude of national government publications (Figure 8.1). De Chiara and Koppelman's book, as a case in point, is a summary of essential standards extracted from many other manuals. It is a sort of dehydrated landscape mix which provides alternative guidelines for lighting, ways of trimming street trees, the design of street signs, house layouts, street classification criteria, parking lots, athletic fields, shopping centre configurations, neighbourhood units, transition zoning, bridge design, and several hundred other elements of landscape.

With the assembly-line preparation of official plans, and the use of well-established models for the design and layout of developments, much urban planning in the post-war period has become little more than a type of planning by numbers: identify the problem or requirement, match to it an appropriate planning device or guideline and develop accordingly. This is, of course, done with varying degrees of sensitivity to local circumstances. Nevertheless, planning by numbers is common enough to have given many modern landscapes a quality of predictable orderliness, a place for everything and everything in its place. Because of it the pedestrian precincts, the arrangements of apartments and houses, the street signs and road patterns, the modern sculptures, all look much the same whether they are in Canberra, Corby New Town in England, or Columbia in Maryland.

There is an irony in all of this. One of the origins of town

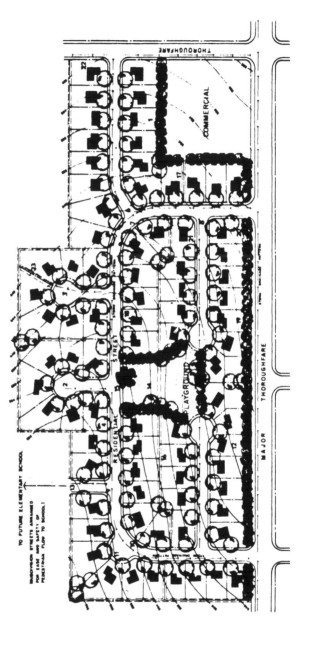

Figure 8.1: Planning by numbers. An illustration from the Community builder's handbook, *1949. The numbered notes refer to numbers on the diagram*

1–52 Subdivision Planning Principles

1. Lots arranged to back onto non-residential use and given additional depth plus planting screen and/or fence along rear lot lines.
2. Cul-de-sacs serve odd parcel of land.
3. Cul-de-sac turn around right-of-way 100′ diameter.
4. Street trees planted approximately 40′ to 50′ apart where no trees exist.
5. Additional building set-back improves appearance of entrance.
6. Street intersections at right angles reduce traffic hazards.
7. Where practical side lot line centered on street end to avoid car lights shining into residences.
8. Three way "tee" intersections reduce hazards.
9. Property lines on 20′ radii at corners.
10. Side lot lines perpendicular or radial to right-of-way lines.
11. Eyebrow increases frontage to produce additional lots in deeper portions of block.
12. Backing lots onto major thoroughfare eliminates hazard of individual driveways entering on it.
13. Provision for access to undeveloped land.
14. Neighborhood park located near center of tract. Adjacent lots wider to allow for extra side line set-back.
15. Pavement shifted within right-of-way to preserve existing tree.
16. Underground power and telephone in rear lot line easements.
17. Ten foot walk easement allows access to shopping adjacent lots wider for additional protective set-back.
18. Variation of building set-back line and orientation along straight street creates interest.
19. Screen planting gives protection from noise and lights of thoroughfare.
20. Intersection planting with foliage above or below driver line of vision.
21. Wider corner lot permits equal building set-back on each street.
22. Lots sided to boundary street where land use across street is non-conforming.
23. Drainage easement to accommodate downhill cul-de-sac.

Source: Urban Land Institute, 1947, p. 175

planning at the beginning of the twentieth century was the reaction against what was called 'by-law planning' — the uniform streets of uniform houses built exactly to the standards of late-nineteenth century building by-laws. In the 30 years after the end of World War II a new by-law planning emerged, scarcely less uniform, and possibly a good deal more comprehensive than the earlier one. Perhaps it has been successful in preventing jerry-building, unscrupulous development, and the worst excesses of advertising and traffic congestion; such preventive successes are, of course, mostly invisible and therefore not easily assessed. It has certainly ensured that social facilities and services like schools and parks have been provided in all new developments. But it has also constrained possibilities and reduced the likelihood of idiosyncratic developments and the 'happy accidents' of juxtaposition (Thomas Sharp's phrase) which make the older parts of cities so interesting visually.

Clean-sweep planning and urban renewal

'Clean-sweep planning' is Alison Ravetz's term for the sort of redevelopment which begins by eradicating whatever exists and then creates something utterly new. It is planning without regard for physical or historical constraints. Clean-sweep planning, which was probably part of every city-builder's dreams since Haussmann had driven avenues through central Paris in the 1850s, emerged as a practicable reality from the destruction of the blitzed cities of Europe. Within a few days of the bombing of their respective cities in 1940 the city fathers and engineers of both Rotterdam and Coventry began to conceive of plans for reconstruction that owed nothing to what had been there before. Not everywhere followed the same path. In central Warsaw, and parts of Cologne and Dresden, the destroyed areas were so meticulously rebuilt as they had been before the bombing that it requires a close effort of observation to know whether the streets and buildings are 40 or 400 years old. Most town planners wanted nothing of this symbolism of historical continuity; they chose instead the modernist symbolism of progress, of transcending the destruction and the past by creating cities better than their predecessors.

The main thing about clean-sweep planning was that there should be as few obstacles as possible to an entirely modern solution. The assumption was that little, and possibly none, of anything old was worth saving or reproducing, and this included not only buildings but also property lines and road patterns. For many European planners the war was seen as a blessing in disguise because it destroyed so many congested and inefficient city districts, and even when buildings had been spared, like the Montague Burton department store in Coventry, they were often expropriated and demolished after the war 'in order to clear the whole area' (Johnson-Marshall, 1966, pp. 178, 304). The whole area could then be replanned in the modern ways. In shopping districts these usually involved pedestrian precincts, and in residential areas they consisted mostly of laying out blocks of apartments and rows of houses in straight lines and heroic geometric arrangements.

Between 1945 and 1960 in the working-class districts of London, in south Amsterdam, at Sarcelles outside Paris, great open-block complexes of apartment residences were built to replace the old row houses and tenements that had been

destroyed in the war or subsequently declared unfit to live in.
The forms of these were not new, but they had never been used
on this sort of scale. Invented by the Bauhaus planners and Le
Corbusier in the late 1920s and early 1930s, these had then been
built only in a few developments in Germany and Holland. Now,
with some modifications, they became the standard for public
housing (Figure 8.2). Often they consisted of several long parallel
rows of three or four-storey walk-up apartments interspersed by
slab blocks and point blocks (the accepted terms for describing
apartment buildings) up to 20 storeys high. Sometimes the blocks
were self-consciously staggered and offset, or otherwise adapted
to their site, and combined with row housing, schools, and single-
storey buildings for the elderly. By 1966 the London County
Council had erected over 100 virtually identical eleven-storey
blocks of low-income housing, mixed with lower rise buildings, in
clusters throughout London (Johnson-Marshall, 1966, p. 236). By
1969 one Parisian in six lived in one of the many *grands ensembles*,
or housing complexes each with 8,000 to 10,000 units, on the
edge of the city (Evenson, 1979, p. 239).

Because of their extent, their height and their lack of sympathy
for the old surrounding geography of streets, these housing projects are highly visible elements of urban landscape. Their
development was often accompanied by a Corbusian type of rhetoric about the advantage of the light and air and open space they
provided even though densities exceeded 100 persons per acre;
and in the architectural models or when seen from the air (and
most of the photos in planning books show them from the air)
these projects did indeed seem to be the harbingers of Le
Corbusier's city of tomorrow.

On the ground they are somewhat less exciting, though in the
best cases, such as the London County Council project built at
Roehampton in the early 1950s, they are not unattractive. Roehampton comprises 15 eleven-storey point blocks arranged in two
groups, six eleven-storey slabs, and several long rows of four-
storey maisonettes and two-storey row houses, carefully arranged
to take advantage of a sloping site and to preserve mature trees
and green open spaces. Most clean-sweep developments were
much less fortunate in their sites, which seem to be mostly flat
and barren. The usual result has been grey, dreary groups of
buildings made of prefabricated concrete sections, offering few
surprises and a superfluity of right angles, the open spaces too
barren and windswept to be of much use for anything. This sort

Figure 8.2: Clean-sweep planning in Europe. Slotervaart in the western suburbs of Amsterdam, from the air and on the ground

Source: aerial view from Benevolo, 1980, p. 904

of thing often replaced a varied and deeply textured urban scene; for example, Evenson notes (1979, p. 263) that in the redevelopment of a slum district in Paris 48 cafés were replaced by one, ten bakeries by two, 49 groceries by five. The drab new surroundings frequently seem to have promoted not the egalitarian, happy city and society of tomorrow but a whole range of personal and social problems including depression, vandalism, difficulties of supervising young children in ground-level playgrounds from upper-floor apartments, and the joyless experience of sharing elevators with gangs of juvenile delinquents.

In North America the absence of war damage delayed the onset of clean-sweep planning, and it was not until the early 1950s that concerns about poor housing and office expansion, coupled with federal government financial incentives, led to large-scale urban renewal projects. While the circumstances were different from Europe the specific and visible results are often not dissimilar — a mixture of high-rise and walk-up apartments arranged geometrically in open blocks.

Urban renewal was seen as a sort of radical surgery to clean out unsafe, unsanitary, overcrowded dwellings which fostered social and economic problems. Public authorities could acquire, clear and replan whole areas using strong powers of expropriation (sometimes called 'condemnation' in the United States) provided by the 1949 Housing Act. The philosophy of the entire process was nicely summarised by the US Supreme Court in a ruling rejecting an appeal against the expropriation of a store in Washington, DC. 'Experts concluded', the ruling reads, 'that if the community were to be healthy, if it were not to revert again to a blighted or a slum area, as though possessed by a congenital disease, the area must be planned as a whole ... It was important to redesign the whole area so as to eliminate the conditions that cause slums' (cited in Futterman, 1961, p. 121). Medical analogies for the city were much in vogue in the 1950s; cities had diseases and it was up to planners to cure them by cutting out the infected parts.

The consequence of this sort of thinking was the bulldozing of large districts of nineteenth-century tenements and row housing, many of them occupied by poor or black communities, deemed unfit for habitation. The old, low-rise, street-oriented houses were replaced by great apartment slabs, surrounded by asphalt and chain link fences, often bordered by expressways or railways (Figure 8.3). Like their European contemporaries, these are distinctive because they often stand out like serried ranks of sore thumbs from whatever surrounding streets happen to have been left.

Clean-sweep planning and urban renewal were founded on

Figure 8.3: Urban renewal in America and England. The Robert Taylor Homes in South Chicago, built in the 1960s, are reputed to be the poorest district in the United States; the building in the foreground is a school. Rationally arranged office blocks in the City of London

the simple philosophy that clean, well-ordered, modern built environments will lead to a healthy and orderly society. Unfortunately social problems are not so easily solved. In the early 1960s urban renewal projects came under escalating criticisms, most notably from Jane Jacobs in *The death and life of great American cities* (1961), who accused them of destroying everything that was vital in urban living. Instead of resolving once and for all the problems of slums and urban decay, they had, she claimed, uprooted communities, scattered them into whatever cheap housing was available elsewhere (thereby spreading the 'infections' of poverty and disease rather than curing them), and they had created their own social problems of disaffection, violence and vandalism.

Other problems also began to emerge. Technically the buildings were energy-consuming monsters, the flat roofs leaked, the elevators broke down and were continually vandalised. Many of the European buildings had been made of prefabricated sections and leaked along the joints of these, and also had problems with dampness, moulds and vermin. In 1968 a point block of prefabricated sections, Ronan Point in London, partially collapsed like a pack of cards after a gas explosion in one unit, and this led to widespread criticism of high-rise blocks in the British press. In fact, the economic merits of such buildings were already looking doubtful and thereafter few were constructed in Britain. In 1972 the Pruitt-Igoe public housing renewal project at St Louis in the United States was partially demolished. This had been designed by the well-known architect Minoru Yamasaki and built in the mid-1950s, and had won a design award from the American Institute of Architects; in less than 20 years it had proved itself to be virtually unlivable. According to Tom Wolfe (1981, pp. 81-3), a public meeting was held to discuss the future of the project and the tenants made the simple suggestion that most of it should be blown up. Since there was no other feasible solution to its problems, it was. Wolfe also describes another urban renewal project, in New Haven in Connecticut, which was built in the early 1970s and had to be demolished in 1981. Since 1975 apartment projects in Liverpool, Manchester and Belfast have also been demolished. With these demolitions clean-sweep, urban renewal planning went into sharp decline.

Megalomaniac projects are still sometimes proposed but fewer and fewer of them are being realised, at least in the Western world. This does not mean that apartment buildings are no longer being constructed, for indeed they are. But now they are less commonly large projects for the poor as stylish point blocks for the very wealthy. Meanwhile the term 'urban renewal' has quietly dropped from planners' vocabulary; they prefer now to talk of 'urban revitalisation'.

Pedestrian precincts, plazas and tunnels

Pedestrian precincts are the product of a specific type of clean-sweep planning originally developed in Coventry and Rotterdam where the central shopping districts had been destroyed by bombing. In Coventry the idea was derived from the Rows at Chester (a medieval pedestrian arcade) and the physical benefits of department stores (Johnson-Marshall, 1966, p. 306). Its actual forms, however, are new. Conceived in 1942, and built in the late 1950s and early 1960s, this precinct is not just a street closed to vehicles, but a series of connected open spaces with stores on two levels connected by balconies. The squares have flowers, pools and sculptures and trees; deliveries are made by back laneways, hidden away and unglamorous. At Rotterdam, as at Coventry, the old property lines and roadways were wiped out (owners were compensated and given the chance to purchase equivalent space in the new development), and the entire blitzed area replanned as a mixture of retailing, apartments and offices, including several conventional shopping streets. The Lijnbaan, the new pedestrian space, was designed as a single architectural entity, and is long, straight and angular. In both Rotterdam and Coventry, and indeed in most pedestrian precincts, the architectural style is a sort of generic or no frills modernism, which means that most of the shapes and spaces are geometric, buildings are often made of precast panels and modular forms, pale greys and browns predominate, the store signs are subdued, nothing is very stylish (Figure 8.4).

Pedestrian precincts were enthusiastically adopted by planners. They were incorporated into designs for the new towns in Britain and Sweden and became a standard feature of redevelopment projects like Saint Paul's Yard in London, La Defense in Paris, and more recently and on a much smaller scale the shopping centre of the original garden city at Letchworth. Traffic-free shopping districts were also adopted in the early 1950s in many European cities, other than those which had been bomb-damaged, by using the simple expedient of closing existing streets to traffic. The narrow streets of medieval cities proved to be especially well suited to such changes since they were too narrow for both vehicles and people; a few movable posts closed the street and could be taken down to permit access for service vehicles in the early morning or at night. In Britain street closings did not begin until the late 1960s, when it became undeniably

Figure 8.4: Different forms of pedestrianisation. The modernist style precinct as constructed in new towns and in the bomb-damaged districts of old cities and not necessarily following former street patterns; this example is in Exeter. The limited vehicle access mall, with widened sidewalks, additional landscaping, and a sinuous right-of-way reserved for buses and taxis; this is the prototype — Nicollet Mall in Minneapolis designed by Lawrence Halprin, opened in 1967, and complemented by a number of pedestrian bridges which link the downtown stores to create an extensive pedestrian system

clear that heavy traffic and the quality of old urban places are inversely related. After experiments in cities such as Norwich and Hereford had proven successful, the pedestrianisation of some shopping streets occurred in most self-respecting towns and cities. Planters with flowers and small trees, kiosks, benches, brick or cobble surfaces, and the bollards that accompany most pedestrianisation projects, have become common features of the British townscape.

The provision of spaces for pedestrians in North American cities followed a very different path. There one of its origins is probably a New York City ordinance of 1961 which permitted a floor area bonus (i.e. allowed more storeys) if the developer provided a publicly accessible plaza at street level. In fact some skyscrapers, notably the Lever House and the Seagram Building on Park Avenue, already included such plazas, and the ordinance was merely trying to encourage their more widespread adoption. Skyscrapers with plazas were soon being built in many North American cities, though most of the plazas were not very inviting — hard, barren, windswept, with a few stunted trees and a piece of sculpture in the middle, they were hardly an improvement on the street canyons they were meant to replace. It was soon realised that, with stores and cafeterias at or below street level, these otherwise little used spaces could both contribute a financial return and be an asset to the city. This may have happened initially in Montreal, where the idea of a network of below-surface, climate-controlled pedestrian streets lined with shops was suggested in the mid-1960s as a way of combining an escape for pedestrians from harsh winter weather with some large-scale new development projects. By applying a system of development bonuses to office construction projects the city planners promoted the construction of tunnel streets and malls on a piece-meal basis until they developed into an extensive subterranean network; this is served by the Metro and links railway stations, hotels and many of the major office buildings and convention centres. Similar systems of underground pedestrian streets have been built in Toronto and Houston. Perhaps following and adapting this lead, it has now become a common practice to include some shops and restaurants on the first two or three floors of office buildings so that these become a traffic-free extension of the street; thus the Citicorp building in New York City has several floors of retailing accessible from street level, but these are not connected into a larger system.

These tunnels and plazas with stores differ from pedestrian-ised streets in one important respect — while streets are public, these are privately owned, controlled by development cor-porations, and subject to whatever rules and regulations the corporations wish to impose, for example, about distributing pamphlets or taking photographs. They also often duplicate the streets outside, threatening their commercial viability by drawing away pedestrians and abandoning them to traffic. This has occurred most obviously in downtown Houston (though there may be other factors at work there) where the skyscraper-lined streets have few stores and most of those without display windows; the central city has turned inward to the underground malls and at street level has become a virtual pedestrian-free zone dominated by parking lots and traffic.

Not until the late 1960s was the possibility of pedestrianising existing streets seriously considered in North America, possibly because by then suburban, climate-controlled shopping malls had begun to take business away from the downtown stores. Though there were some temporary, experimental street closings in the heady days of the late 1960s, and isolated permanent cases like Sparks Street Mall in Ottawa, it was probably Nicollet Mall in Minneapolis, designed by the landscape architect Lawrence Halprin and constructed in 1967, which initiated their wide-spread adoption. Nicollet Mall is not completely closed to traffic. A single, sinuous traffic lane for buses, taxis and emergency vehicles was left open, and the sidewalks were widened and land-scaped. In the harsh winters of Minnesota this alone would have not been enough, so Nicollet Mall was combined with a system of bridges linking department stores and office buildings to provide interior access to much of the downtown commercial area. This sort of complex interior/exterior, public/private pedestrianis-ation, or partial vehicle access pedestrianisation, has become a common feature of the revitalised centres of North American cities in the 1980s. Examples are to be found in Niagara Falls in New York State, Calgary, Vancouver, Quebec City, New London in Connecticut, and in both Georgetown in Washington DC and San Antonio alongside their renovated canals.

State and corporate new towns

There are two distinct types of modern city planning. Planning in

city centres has to cope with established street patterns, buildings, communities and vested interests, and is thus a politically
saturated activity; planning in the city fringe or for outlying sites
deals with the replacement of unpeopled countryside by built
environments and is largely technical and apolitical. In Britain
the term 'greenfield development' is sometimes used to describe
developments where there is no resident population or interest
group to protest, and the planners and developers can simply do
whatever they think is best and without compromise. New towns
are the purest cases of late twentieth-century greenfield developments.

It is conventional to treat new towns as entirely distinctive
from suburban developments, largely on the grounds that in
Europe new towns are state ventures whereas suburbs are privately built. The fact is, however, that post-war suburbs and new
towns exhibit many landscape similarities partly because they
were contemporary and employed the architectural fashions of
the day, partly because they drew on a single set of planning concepts, and partly because they were subject to the same building
standards. This is particularly so in North America where the
new towns have been built by the same sorts of corporations
engaged in suburban developments. A purist might have it that
these North American cases are not real new towns (see, for
example, Clawson and Hall, 1973, p. 198ff). Another inter-
pretation suggests that a new town is any substantial settlement
planned in its entirety and offering local employment (see, for
example, Osborn and Whittick, 1977, pp. 98-9). Even this latter
definition is too constraining for some developers who happily
describe any large residential development as a new town,
regardless of employment opportunities. Rather than imposing
arbitrary restrictions on the definition I prefer to include all of
these ideas but to distinguish state and corporate new towns.

Though the inspiration for new towns came from the garden
city movement, the construction of them was promoted chiefly by
the British New Towns Act of 1946, which provided the necessary
mechanisms for the construction of 14 largely autonomous towns
to take up the overspill population from major cities, the growth
of which was to be restricted. Subsequently a further 14 towns
were designated and constructed, several of them based on existing settlements, and two — Telford and Milton Keynes — integrating several villages and towns into regional complexes with
populations of more than 200,000. In addition, about 50 existing

towns underwent planned expansions to accommodate overspill from the cities. By 1976 about 1,800,000 people in Britain were living in new towns, and several hundred thousand more in expanded towns, which were old places expanded along new town lines. Similar programmes of old town expansion and new town construction were adopted in Sweden, Russia, Finland, Holland and France. According to Osborn and Whittick (1977, pp. 467-90) new towns have been built in more than 75 countries since 1945.

The state planned towns of Europe are based on the ideal of providing a mix of social and economic classes, an employment basis, and equal access for all to the new facilities. In contrast, corporate developments, whether suburban estates or new towns like Reston in Virginia or Kanata near Ottawa, are planned and packaged chiefly to maximise consumer satisfaction and company profits, and many of them are designed to take advantage of existing employment opportunities in nearby cities.

These differences in intention and ideology have frequently been magnified by adopting ideologically appropriate styles of building. In Europe state new towns and housing developments were promoted especially by socialist governments, and architects and planners almost without exception opted for generic modernist styles of building and planning. Apartments and row housing, with their close-packed modular living units, serve to create relatively high residential densities and to symbolise egalitarianism. Community provisions are made for all age and income groups, and include day-care centres, apartments for the elderly, playgrounds and pedestrian precincts. In contrast the privately planned development is distinguished by the prominence given to cars and to single-family detached houses with their own gardens and garages. The styles of these invariably derive from vernacular and traditional designs — fragments of georgian, tudor, and colonial are especially common. There are shopping centres and malls, not pedestrian precincts, and social facilities are fewer and often seem to be a bit out of the way, as though provided grudgingly to comply with planning requirements.

State planners usually look upon corporate developments with considerable contempt. Nevertheless the fact remains that they are thoroughly planned according to much the same planning principles as the state-sponsored towns. Columbia New Town, for example, half-way between Baltimore and Washington, DC, is the inspiration of one business magnate — James Rouse (a promotional movie for Columbia states simply that 'It was always his dream to build a city'). His ultimate goal is, of course, profit, but Columbia is also a model for American residential planning. It is to have a final population of 60,000, with employment opportunities for 37,000, mainly in hi-tech industries. About 35 per cent of the area acquired for the town will remain as open space. Columbia is arranged hierarchically into eight 'villages' with their own community centres with recreation facilities and stores, each of which is further sub-divided into two or three neighbourhoods of 2-3,000, focused on the elementary school. Most housing is fully detached but there are also row houses and small, low-rise apartment buildings; there is one district of very expensive houses called, for no apparent local reason, Hobbit's Glen. Housing is arranged so that the highest densities are adjacent to the community centres. At the centre of the town, on the side of an artificial lake, there is a commercial area with a mall, offices, hotels, the main library and high school. This area has an enclosed shopping mall, and a system of pedestrian bridges and walkways so that traffic and pedestrians are separated. As with every new town I have visited, state or corporate, this main centre is divided from the residential areas by roads and open space of uncertain purpose. Strict controls are maintained over all residential and commercial developments; the gas stations are so discreetly hidden away that one has to know where they are in order to find them.

Columbia uses similar organising principles of residential and retail organisation to those employed in British and Swedish new towns, though it realises them somewhat differently. Compare it with Corby, a new town in central England designated in 1950 and expanded in the 1960s to accommodate a population of about 55,000 (Osborn and Whittick, 1977, pp. 331-43). The town is divided into 13 neighbourhoods with populations ranging from 1,500 to 13,000, and these are further sub-divided into housing clusters of 300 to 400 dwellings each. The smaller neighbourhoods have one primary school, the larger ones have two; most of them have neighbourhood centres of shops and community facilities. There are three grades of housing — flats and maisonettes, two-storey row and semi-detached houses, and relatively low-density owner-occupied detached houses built by private developers. The original town centre was made into a pedestrian precinct in the 1960's expansion; it, like the centre at Columbia, is surrounded by car parks and ring roads. A final note: the major industry at Corby was a steelworks which, against most previous expectations, was closed in the early 1980s as part of a programme to rationalise the British steel industry, and the town now has a poor economic base and high unemployment.

In the two decades after World War II the planning of new towns carried with it many of the socialist hopes and utopian visions that had inspired the earlier garden city movement. New towns offered a solution to the problems of decaying cities and a model for a new type of urban society. In practice, however, planners mostly fell back on a repertoire of conventional concepts and devices, the same devices that have been used in all post-war residential planning — neighbourhood units and housing clusters, hierarchies of roads and shopping facilities and parks, mixtures of housing types with highest densities near the community centres, and networks of pedestrian paths. As a result new towns have mostly turned out to be particularly fine examples of planning by numbers. From within, and from what can be seen on the ground, it is difficult to tell whether one is in a new town, an expanded town or a new suburb of an old town, except perhaps at the edges of the distinctively isolated new town shopping centres. Indeed in the latest generation of English new towns, notably Telford and Milton Keynes, it is difficult to tell whether you are in a town at all. Laid out on a loose modular grid of one kilometre squares, with roundabouts (traffic circles) at the intersections, these were apparently designed for a mobile modern society. It might be very different to live there, but for the visitor the overwhelming impression is of road and roundabout, road and roundabout, a maze made exclusively for motorists in an interminable urban fringe.

With declining population growth coupled with an increasingly uncertain economic future the new town movement has since 1970 almost everywhere fallen on hard times. The last British new town, Milton Keynes, though still being built, was designated in 1967, and in North America there is no current equivalent to Columbia. The most recent organisation of new towns are usually resort or retirement communities, such as La Grande Motte on the Mediterranean coast of France or Lake Havasu City in Arizona; or they are just larger than normal suburban developments, like Audubon New Town outside Buffalo. These are hardly the autonomous settlements envisaged by planners in 1946. It appears that, like clean-sweep urban

157

renewal, the planning of new towns is an idea whose time has passed.

Highway design and the demise of the street

Since 1945 highway planning and engineering have greatly expanded a landscape that had begun to be made in the 1930s, a hard-wearing and relatively featureless landscape oriented exclusively to machines and machine speeds. Its forms are so familiar and extensive that their newness is difficult to hold in mind — post-war additions to the view of and from the road include almost all expressways and freeways, countless arterial roads and road widenings, most filling stations, road markings, parking lots, arrays of parking meters, and the No Parking and other traffic signs which have sprouted like metallic mistletoe on every available post.

One of the more insidious results of all of this has been the demise of the street. We can tell from the mixed-use streets that still remain that they can serve a variety of formal and informal community needs, including shopping, parades, strolling, traffic access, demonstrations, and chance meetings with friends. In planned developments, and that includes just about everything built since 1945, there really are no streets; they have been redesigned as any one of a number of types of roads: collectors, distributors, arterials, by-passes, highways, motorways, inner ring roads, relief roads, expressways. At the same time pedestrian precincts and paths, malls and plazas have been built to accommodate all the former street activities of pedestrians. The new roads are perhaps not used by less people than streets, but they are used for less activities and really just for parking and moving vehicles.

The appearance of these new roads is largely a function of the engineering requirements necessary to accommodate vehicles moving in different quantities and at different speeds. The greater and faster the traffic the wider the road. For each category of road — local, collector, arterial, expressway, and so on — precise design standards have been developed, which, since they are determined by the size and speed of cars and trucks, are much the same internationally. The lane widths, curvatures, gradients, intersection designs, lighting, signs, materials, crash barriers, have a safe predictability to them (Figure 8.5). They are made of

Figure 8.5: The mechanical road. Cross-section standards for different types of roads, as presented in a Canadian public works manual

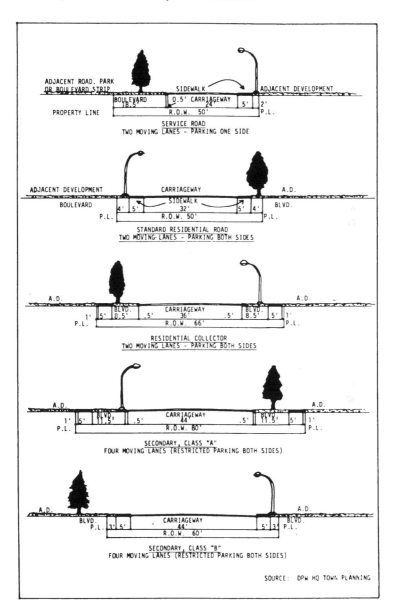

Source: Schwilgin, 1973, p. 87

the universal materials of concrete, asphalt and metal, and because they have to be seen at speeds of up to 100 kilometres per hour their forms must be without distracting detail. There are some regional differences in design and in landscaping, for instance many of the freeways of southern California are lined with lush vegetation and have medians planted with flowers (these have been appropriately described as 'untouchable parks'), but in most places they have trimmed grass with a few trees and it is only the licence plates and the names on the road signs which tell you that you are driving on an expressway around Antwerp rather than Houston, or Melbourne rather than Birmingham.

The construction of new roads and the reconstruction of old roads to these rigidly applied standards has fundamentally altered the appearance and the form of cities. In detail it is all too common to find roads of the standard width, with standard lights at regulation distances, standard curbs and sidewalks, that have been forced through on the line of an older road without apparent regard for existing development. Indeed I know of remote hamlets in South Wales where new clusters of three or four suburban-style houses are provided with the package of street lights, sidewalks and curbs, and the old roads are left thankfully in unkempt darkness. There is scarcely a harder task in protesting a development than getting highway engineers to relent on design standards.

At a larger scale the web of expressways, inner loops and ring roads, by-passes and improved urban throughways have created a new urban pattern that the historian Sam Bass Warner (1972, p. 46) calls 'the wheel' — a pattern of circumferential and radial roads which has been superimposed on to existing city forms and by which new growth has been shaped. Like igneous intrusions in sedimentary rocks these cut across the grain of the city with their distinctive landscape textures. Their professed purpose is to alleviate congestion on local roads while improving access from the suburbs to central districts. The consequence is that in the clearest cases, and Bristol and Birmingham in England are unfortunately good examples, the city centres have virtually disappeared under an onslaught of road improvements and widenings, shopping and office areas are encircled and severed by four-lane roads, the remaining sidewalks have had to be separated from roadways by metal safety fences, and pedestrian crossings replaced by tunnels.

The circumference of this wheel of expressways has, since

about 1975, begun to generate a pattern of new commercial and industrial centres, variously called suburban downtowns and urban villages, though they are neither downtown nor remotely like villages. These are usually located at interchanges and consist of clusters of sleek offices and factories, motels (about 150 small office and industrial buildings can support a 250-room hotel), restaurants, and perhaps a shopping mall, local government offices and condominium apartments, all surrounded by land-scaped strips and by parking lots which are, in many cases, as large in area as the floor space of the buildings. Most large metro-politan areas have several more or less clearly defined urban villages; sometimes they are strongly nucleated; sometimes they are strung out along the expressways with the corporate signs and logos carefully oriented like billboards to the passing traffic; a distinctive version of them is to be found near every international airport. They are in part a reaction against urban sprawl and an attempt to provide some sort of employment focus in suburban areas, in part the products of large-scale corporate developments integrating several land uses, and in part a response to tele-communications, a service economy and automobile commuting (Leinberger and Lockwood, 1986, p. 45). Electronics and data processing firms such as Digital, IBM and Apple, and companies which can maintain electronic contacts between their offices, seem to be especially common. For surburban commuters these urban villages are especially convenient because they are close to home. They have, however, begun to generate their own traffic problems. Traffic jams on expressways are now two-way rather than one-way heading into or out of downtown.

Attempts to prevent this sort of fate for all cities were made in the 1960s. In Britain a major government study, the Buchanan report, published in 1963, recommended accommodating the automobile by a variety of different strategies depending on local circumstance, rather than universal capitulation to it through the construction of more and larger roads. At about the same time in the United States the planner Victor Gruen proposed a different solution — a loop expressway around downtown, with adjacent parking facilities, would serve to divert and store traffic and to protect the heart of the city, which could then become an area mostly for pedestrians. Gruen's specific idea was not widely adopted, Rochester in New York State may be the only city to use more than fragments of it, but since the mid-1960s there has been an ongoing confrontation between those who would have more automobile access to city centres and those who would prevent this. The victories of the protestors are sometimes to be seen in expressways which lead nowhere, like the Embarcadero expressway in San Francisco which ends literally in mid-air, or the Spadina Expressway in Toronto which stops far short of its intended destination downtown, was renamed the Allen Road, and now filters off into local streets.

The victories of the engineers are much more obvious; they are apparent in all the expressways penetrating, weaving, looping around almost every city in the developed world. Wherever possible these have been driven along the lines of least resistance — parks, river valleys, blighted areas with low property values, or raised on columns above the rooftops of old buildings. They are ambiguous things. Looked at in detail they are almost always ugly and dirty, like the railways before them, they divide up cities, separating industrial districts from residential areas, or poor ghettos from rich communities. But they have also made the enormous scale of twentieth-century cities comprehensible by making all districts accessible from anywhere within minutes. And they have created an entirely new experience of urban landscapes, an experience comprised chiefly of sequences of concrete channels and glimpses of landmarks, with sometimes dramatic views of skyscraper skylines unfolding as one drives toward them or around them. This latter fact is sometimes exploited by designers. The architect Philip Johnson says he designed the twin towers of the Pennzoil Building in Houston so that the narrow cleft between them would appear to drivers on the Inner Loop Expressway as a brief and magical flash of light.

How to recognise a planned place

The landscape effects of planning should be obvious. Nearly every urban development in North America and Europe since about 1950 has been made subject to planning control, so anywhere that is new is almost certainly planned. Things are, however, not quite this simple. There are those, like the environmentalist Paul Ehrlich, who, being unfamiliar with recent urban planning, look at sprawling suburban developments or expressway patterns and criticise them as being chaotic and unplanned. This criticism is simply wrong; these environments are precisely planned and he is expressing his personal dislike for what he has failed to recognise as a planned development. But his mistake does have some justification because the visible evidence of planning is sometimes so obscure that it evades even careful observation. Consider SLOIP. SLOIP is Lionel Brett's expressive acronym for Spaces Left Over in Planning, usually empty fields and ill-maintained lots awaiting development. In new towns, for example, development is phased, so that land has to be left vacant while population growth catches up with the projections. In 1981 the town centre at the new town of Milton Keynes was still partially surrounded by what appeared to be abandoned farmland but was shown on plans as 'reserved sites' (Figure 8.6). That was true SLOIP.

Figure 8.6: SLOIP (Spaces Left Over In Planning) on either side of a mechanical road and adjacent to the commercial centre, seen in the distance, of the new town of Milton Keynes in England. Presumably this space is intended for expansion of the centre, but in 1981 it was rough grassland

Even in areas which have been built up there can be messy results. If one stands at the intersection of two arterial roads and contemplates the wires, the poles, the variety of signs, the spaces, the bits of vegetation, the buildings, all of them apparently put together into arrangements which could not be worsened by deliberate design, it is indeed difficult to believe that any of it has been planned. Yet at the very least the set-backs, the land uses, and all the roadway standards will have been made subject to planning approval. It seems that most of the efforts of planning to create order are concentrated in the heart of land use zones. Where these zones meet, usually along roads, matters seem to get out of control, so that the whole scene somehow turns out to be less than the sum of its parts.

Firmer evidence of planning is to be found in any new landscape with straight lines, geometric layouts, curvilinear streets, anything arranged in rows from sidewalk benches to apartment slab blocks, and anything with neat edges and parallel lines, such as curbs and sidewalks (Figure 8.7). Neither nature nor unplanned humanity are in the habit of arranging themselves and their possessions in neat rows. Of course, renaissance and georgian builders also employed geometric forms in laying out towns, as in Edinburgh, Karlsruhe and Bath, but usually this was done with a great confidence and sense of proportion. The

Figure 8.7: A planned place in the segregated city. A standardised distributor road, lined by houses at the uniform set-back prescribed in zoning by-laws; in the middle distance the horizontal roof-line of the community shopping centre can just be seen; and behind that an array of apartments in the high-density zone. An unremarkable and tidy patchwork of functions in suburban Toronto

geometries of modern planned landscapes are, for reasons I cannot quite identify but might have to do with poor proportions and inconsequential decoration, unsatisfying; some of the arrangements are perhaps too obvious, others have incomplete and overlapping forms with lines of perspective broken or unfocused. Half-hearted geometries are a clear indication of modern planned environments.

Finally there is the segregation of activities, nothing overlapping or overflowing at the edges, no mixing or confusions. This is an indication of the obsession with orderliness that grips all modern planning, though perhaps less so now than before 1975. Look at almost any planned place, Jane Jacobs has written (1961, p. 447), specifically with reference to a small riverside park which got 'improved': 'An all too familiar sort of mind is obviously at work here, a mind seeing only disorder where a most intricate and unique order exists; the same kind of mind that sees only disorder in the life of city streets, and itches to erase, standardize, suburbanize it.' The entire planning process, she claims, consists in trying to obtain 'decontaminated sortings' (p. 25). She may in fact be describing a common cast of modern rationalistic thinking, but there is no question that it has been exceptionally well represented in planning. Evidence can be found in most new urban landscapes: the arterial roads that separate business, industrial and residential districts, the fences that divide zones of single-family houses from apartment complexes and apartments from stores, the neutral spaces in shopping precincts between clusters of benches and flower-beds. Each activity has been assigned its territory, and the lines on the land use zoning maps in the planning office have been faithfully reproduced in landscapes as barriers or neutral spaces. The overall result is an urban landscape characterised above all by its tidy patchwork of functions, a place for everything and everything in its place. There can hardly be a better summary of it than the one by Alison Ravetz; she calls it 'the segregated city'.

9

The Corporatisation of Cities: 1945-

The lives of the poor and the powerless are directly affected by landscapes, for there is little they can do to alter their surroundings and have to adapt to them no matter what they are like. The wealthy and the politically powerful, however, are able to alter landscapes on a grand scale to meet their needs, to express their social superiority and to make more money. In modern society there are none richer and more powerful than corporations.

The french term for a corporation is 'société anonyme'. This reveals far more than the usual definition, which is that a corporation is a large business, publicly owned by shareholders and run by a board of directors. The fact is that it matters little to most of us whether some great company is publicly or privately owned because they all, in spite of the familiarity of their brand-name products, seem to be run by faceless minions whose only concern is to increase productivity and profit. And they have been so successful in doing just this that an ever smaller number of increasingly large multinational corporations have become the primary centres of wealth, power and technical expertise in modern society. According to the economist J.K. Galbraith (1968, p. 21), the entire worldwide industrial system now consists of 'a few hundred technically dynamic, massively capitalised, and highly organised corporations'. Anyone who has travelled in more than one continent will have guessed as much already.

Ours is therefore a corporate world, and corporatisation has become one of the major forces in urban landscape change. Corporatisation is the take-over by business corporations of some previously small-scale or state-run operation. It began in earnest in the nineteenth century with the construction of the railways and the associated hotels and workers' housing, it rapidly became

166

more pronounced in the early decades of this century as companies like Bell and Woolworths insinuated themselves into the fabric of towns, and since 1945 it has been comparable to urban planning in the scale and internationalism of its effects. Indeed, corporations almost seem to be engaged with planning departments in a great dialectical process of confrontation, compromise and construction. Planning regulations unquestionably condition the appearance and layout of late twentieth-century cities, but the content of these and their dominant new visual facts — the skyscraper offices, subdivisions, hotels, condominium apartments, theme parks, shopping centres, retail outlets, billboards, and artworks — are owned, financed, developed, constructed and maintained by corporations.

Manifestations of corporatisation

If something in an urban landscape is tall or wide, if it has looming presence, if it is flashy or colourful, it is almost certainly a corporate product. Governments are also inclined to do things to landscapes on a big scale, but have recently chosen expressionist styles for city halls and other special buildings, and these are rarely taller than the new corporate offices around them. Lesser government buildings are marked mostly by their lack of distinctiveness, a sort of grey, homely, standardised plainness. In contrast, corporations have almost without exception chosen modernist skyscrapers for their downtown offices, while their lesser buildings, the stores and outlets, have been distinguished by flamboyant designs and colourful signs.

Actually this apparently straightforward distinction between the landscapes of business and government does not stand close scrutiny. The fact is that even the government buildings have been constructed by contracting corporations, and may well have been built on land provided by development corporations at low cost as a gesture of apparent magnanimity. The United Nations Plaza in New York City was built on a site acquired by William Zeckendorf, one of North America's leading post-war developers, and then sold to the United Nations (via the Rockefeller family) for whatever they chose to pay (Marriott, 1967, pp. 231-2). Of course, when the United Nations took the site the values of adjacent properties, also owned by Zeckendorf, soared. At a somewhat humbler level, the Scarborough Civic Centre, the city hall complex for a suburban municipality of Metropolitan Toronto, was built on land offered to the municipality by Trizec Corporation, a company controlled at that time by Eagle Star Insurance of London which conveniently also had a controlling interest in the adjacent shopping mall.

These sorts of developments are the most anonymous of all forms of corporatisation. At Scarborough Civic Centre the only visible indication of Trizec's involvement is a short access road called 'Trizec Gate', a name which means nothing to the uninitiated. But corporatisation reveals itself in a whole spectrum of manifestations from anonymity to blatancy, from a hidden presence in land assembly, to a studied presence in office towers, to a highly visible but temporary presence in corporate suburbs. It is only inside shopping malls, along commercial strips, and in urban centres like Times Square and Piccadilly Circus, that the corporate domination of the modern economy comes leaping out at us, almost screaming for recognition and consumption.

Towers of conspicuous administration

No feature of the modern city is more striking than its skyline of skyscraper offices. Manhattan had achieved a version of this by the 1920s, and there were by then suggestions of it in other North American cities such as Chicago, but it is only since about 1960 that almost every prosperous city in the world has acquired a clutch of new office towers and the resulting corporate skyline.

By their great size office towers indicate wealth and economic power. They are the cathedrals of the modern city. They are also corporate flagships, often displaying the company name or logo on high, occupying prominent city centre locations, striving for architectural prestige within the narrow realm of high-rise modernism, frequently designed by famous architects, and invariably a few storeys taller than the already completed buildings of their business competitors. Lever House, the first post-war glass-over–steel-frame building in New York City, is a gigantic advertisement for making 'cleanliness commonplace', as it states on a plaque at street level; not only is it shaped like a great detergent box but its glass surfaces are washed by the same products Lever manufactures for domestic use (Reynolds, 1984, p. 151). When conditions are entirely propitious a corporate tower can become an urban landmark, a subtle and prestigious form of advertising; the pyramidal Transamerica Building in San Francisco is an outstanding example, or on a smaller scale there is Shell House on the south bank of the Thames in London. These are the exceptions, however. The more common pattern is for individual buildings to merge into an architectural mass of similar towers, with the entire scene conveying the direct and forceful message of economic prosperity through concerted corporate investment and administration (Figure 9.1).

Figure 9.1: Capitalise: the corporate skyline of the modern city declares its ideological underpinnings. Houston, Texas

This message is most easily absorbed at a distance, especially from an expressway looping around the business district. At street level the chief impression is one of monolithic austerity. Close up the towers invariably look as though they have been enlarged directly from the architects' models, so the larger they are the less detail they have. They have often taken over whole city blocks formerly occupied by a multitude of different shops, replacing detailed architectural textures and on-street vitality with great blank façades and deserted plazas (Figure 9.2). This elimination of texture and variety from city streets has proceeded in a lock-step fashion with the change from mostly small to mostly big business that has taken place in capitalist economies in the past hundred years, and particularly since 1945.

Figure 9.2: Changes in the street texture because of corporate development; from independent stores with varied façades to entire blocks taken up by a single building. This visible change has been accompanied by a shift of activity from outside on the street to inside in offices and enclosed shopping malls

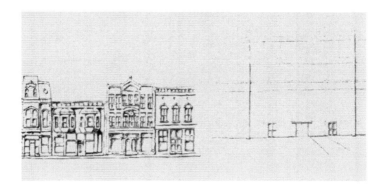

Modernist architecture does not lend itself to gaudy displays and embellishments, so at street level corporate names and addresses are inconspicuously displayed in tasteful metal signs rather than the corporate colours. The word 'skyscraper' is universally disavowed, and the offices are called towers (the John Hancock Tower), buildings (the Seagram Building), or, rather strangely, houses (BP House), or, more recently, centres (the Renaissance Center). To conform with planning by-laws part of the site often has to be left as a plaza. The design quality of these varies enormously but a common pattern is a rectangular area of concrete slabs with a few rectangular flower beds and a sculpture of the unadorned, low maintenance, modern sort — perhaps a set of metal shapes poised at impossible angles, or a group of clumsy bronze figures, or, in rare instances, something whimsical like Picasso's great beast in front of a tower in Chicago (Figure 9.3).

Corporate towers are entered through revolving doors. There is usually little to see on the ground floors, just a security desk, perhaps a mural, a list of all the tenant companies, and a battery of elevators to transport executives to the hidden nerve centres of the corporations above. There are no products on display or services advertised. Here corporate wealth is revealed discreetly through expensive finishes of chrome and polished Italian marble. No doubt the installation of these requires considerable craftsmanship; from what can be seen it is a craftsmanship exercised without responsibility or joy, a craftsmanship that follows blueprints exactly. Office towers may be the modern equivalents of cathedrals but individual workmen leave little evidence of their skills in them.

Figure 9.3: A modernist skyscraper plaza plus sculpture. Actually an exception to prove the rule, this is Daley Plaza in front of Chicago City Hall, a rare skyscraper city hall. By its appearance this ought to be a corporate office, only the large size of the plaza suggests that it is not. The sculpture is by Picasso

Levittown and the corporate suburb

Since World War II residential development has been almost completely taken over by corporations. In Britain in 1930, for example, 84 per cent of the contracting firms building houses had less than ten employees and built only a few houses each year, a fact which ensured some variety even when most of the contractors were using similar pattern books (Oliver, 1981, p. 98). In contrast, in 1971 one company, Wimpey, built 10,200 of the 55,000 industrialised dwellings constructed for local authorities in Britain (Crosby, 1973, p. 103). Similar trends occurred elsewhere. In Canada, for instance, in the late 1940s 80 per cent of all houses were built by companies with less than 20 employees constructing only 15 to 20 units per year. By 1970 80 per cent of all houses were being constructed by large corporations such as Wimpey and Monarch Homes, each building several thousand units a year and using assembly line techniques involving prefabricated roof trusses and stairs, and specialised teams of workmen moving from site to site in sequence. The result of this corporatisation was, especially at first, a dreary uniformity of house styles and layouts widely condemned as a 'blandscape' and occupied by scarcely less uniform white middle-class families.

About 75 per cent of the housing stock in North America has been built since 1945, much of it in corporate built suburban tracts. The prototypes for these, no doubt because they received so much media attention, were the Levittowns of Long Island, Pennsylvania and New Jersey (Gans, 1967). In the years immediately after the war William Levitt, the owner of a small contracting firm which built surburban houses, took advantage of the dramatic demand for inexpensive houses caused by the return of servicemen. For the Long Island Levittown, begun in 1947 for a total of 2,000 houses, he developed an assembly line construction system using concrete foundation slabs, precut sections and a single house style, the Cape Cod, with some façade variations to provide individuality. To meet a continuing demand the Levitts acquired more land, and by the end of 1948 had completed 6,000 houses. These were not laid out according to any master plan, but, possibly following federal housing administration guidelines established in the 1930s, they were grouped around 'village greens' with neighbourhood shops, schools, playgrounds and swimming pools.

A similar *ad hoc* approach of adding neighbourhoods one by

one was used at Levittown in Pennsylvania, begun in 1951 (Figure 9.4). This development eventually accommodated 67,000 people in 17,000 houses. Again they were grouped in neighbourhood units, but here most of the local shops were replaced by a single large plaza on the edge of the tract because in the Long Island Levittown residents had shopped at a regional plaza, not owned by the Levitts, rather than the local stores. Three house styles, Cape Cod, Rancher and Colonial, were used to provide greater variety of choice and improved street appearance.

The Levittowns were built in the context of federal guidelines which advocated curvilinear streets, neighbourhood units and some variety in façades, though these did not have to be followed by the developer. Local planning regulations were an insignificant influence. Most of the planning and design, including the provision of community facilities, were done by the Levitts who had no interest in providing an employment base or creating politically autonomous settlements. The Levittowns were large-

Figure 9.4: Levittown in Pennsylvania in the early 1950s. This is one of several Levittowns which served as prototypes for many of the corporate suburbs that have since been built

Source: Lynes, 1949, p. 270

scale, middle-class, white (only with anti-discrimination legislation were blacks allowed to buy), dormitory suburbs. All of
which was probably to the satisfaction of most of the buyers;
Gans (1967, pp. 146-7) observes simply that what they wanted
was a new house not a new life.

Most suburban corporate developments in the 1950s were on a
much smaller scale than the Levittowns, occurring usually on
blocks of about 160 acres, just large enough for one neighbourhood (Harvey, 1984, p. 9). Not until the 1960s did large-scale corporate suburban developments become common. By then a
rather different demand for suburban houses had emerged, one
which stressed style and variety. The more recent developments,
even when undertaken by a single company, have been divided
up and packaged as distinctive 'communities'. The identities of
these developers' communities are mostly fabricated for marketing purposes; the only thing that the individual houses have in
common is location and a synthetic personality and name. The
names usually suggest something bucolic, romantic or nostalgic
— a sample from a real estate section in the *New York Times* in
1983 includes Greenbriar, Hastings Landing, The Mill, Glen
Oaks Village, Rustic Woods, and Heritage Homes of Westchester
where Quality is Second Nature. There is nothing remarkable
about these, similar pseudo-community names are used everywhere for new residential developments. The relationship
between them and the actual landscape is quite arbitrary, not
least because the site has generally been stripped bare to permit
the easy installation of sewers and roads. During the construction
and marketing phase of development there is a strong corporate
presence in the landscape. Corporate names are prominently displayed on construction equipment and outside display homes
because each corporation believes it has established a solid repu-
tation for dependability and sound workmanship; thus Wimpey
does not just build houses, it builds 'Wimpey Homes'. Of course
all these signs and names are just marketing devices; they are
removed and forgotten as soon as all the houses have been sold,
and the fact of corporate production of the landscape soon fades
into the past.

Since the late 1940s the houses in corporate suburbs have
become larger and more distinctively stylish. At first, as at Levittown, they were usually small boxes without garages, and with
only the most rudimentary detailing. By the 1960s in North
America sprawling ranchers and backsplits with attached garages

had become common; these still tended to be rather graceless and had few distinctive style characteristics. At the beginning of the 1970s houses began to assume clear traditional details, such as fake half-timbering and colonial/georgian doors and windows. By the mid-1980s full-scale revivals of victorian, colonial, neo-classical and other styles, with appropriate elevations and decorative brickwork and trim, had become fashionable (Figure 9.5). Whole subdivisions might be given a consistent style treatment, though no adjacent houses ever have the same façade. In the most expensive developments the façades of century-old houses are now being accurately duplicated, with the two- and three-car garages being neatly hidden around the side.

The growth in size of development corporations since 1945 has also made possible the construction of apartment buildings, the financing and construction of which is simply beyond small firms. Because of their sheer bulk these make a prominent con-

Figure 9.5: Suburban houses since 1946 have become progressively larger and more stylised as lot sizes have become smaller. a) North American Cape Cod style houses of the Levittown period, about 1950; b) asbestos prefabs, Newport, Wales, built about 1950; c) a North American rancher, 1950s and 1960s; d) individualised façades for townhouses, Columbia, Maryland, 1970s; e) detached houses with leaded windows and vaguely classical porches, Elstree, England, 1970s; f) Victorian revival, with decorative quoins and brickwork, and prominent garage, Toronto, 1980s

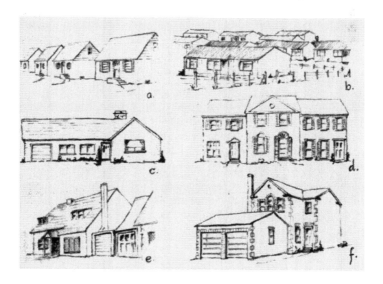

tribution to urban landscapes, whether free-standing or in clusters. This same bulk, especially when under the shadow of modernism, also limits the possible variety of designs. There are point blocks with a square plan, great slabs with an oblong plan, and modifications of crosses and Y plans (Figure 9.6). Some differences in façades are achieved by changing the details of balconies, but have these rarely sufficed to give much individuality to apartment blocks.

Up until about 1975 these look-alike apartment buildings were considered a second-rate form of housing, most of them were for public housing or rental, and were located in less desirable parts of cities. Since then there has been a change in their social status accompanied by a change in their design. Many of the newer apartments are luxury condominiums, very expensive and equipped with the paraphernalia of modern affluent living — exercise clubs, solariums, squash and tennis courts, swimming pools, and elaborate security systems. These command the best sites in city centres, overlooking the ocean front in Miami and Los Angeles, the lake fronts of Chicago and Toronto, the hilltops of San Francisco. Close up they are immaculately maintained, their designs are sleek and sophisticated, but from a distance they are virtually indistinguishable from the basic point blocks and slab blocks of their poor cousins.

In marketing suburban houses and new condominium apartments the development corporations are trying to sell a consumer product, and in this they have to be sensitive to changing demands. So this evolution towards increasingly stylishness is partly a reflection of shifts in popular fashion and taste, coupled perhaps with a general increase in what people can afford to pay for a place to live. It is significant that the earlier designs of houses allowed scope for do-it-yourself additions, out of which a sort of vernacular variety of garages, porches and shutters has subsequently emerged. By comparison the most recently built corporate houses, and of course apartments, are totally designed packages, customised in some details such as the colour of the bricks perhaps, but otherwise completely finished down to the coachlamps and co-ordinated colour schemes. The fact that these houses cannot be easily modified on the exterior seems to be to the modern taste since many new home buyers immediately add to their house packages by acquiring a landscaping package for the front yard, a layout of plants and interlocking brick sidewalks designed by a landscaping company. The idea seems to be that

Figure 9.6: Apartment buildings since 1950 have been built in a limited number of basic forms. Since about 1970 these have become more stylish as apartments have become popular among upper income groups. The old, basic forms are usually located in renewal projects and suburban clusters; the new, expensive condominium apartments stand alone in city centres, on lake-fronts and ocean fronts. a) Point blocks (with a square plan), Amsterdam; b) A slab block, Brussels; c) a Y-plan tower, Toronto; d) a condominium apartment tower with glassed-in balconies and galleria entrance, downtown Toronto. The photograph is of ocean front apartments at Marina del Rey, Los Angeles

everything on the property must bear the stylish look of professional design. Individuality in the corporate suburb is now expressed not by making substantial modifications and additions to one's house, or even by adding idiosyncratic lawn ornaments or a set of shutters; individuality in the houses of the 1980s is expressed by one's choice of completely finished corporate products.

Corporate malls

New suburban house owners need to buy things to furnish and maintain their new houses, and being sensitive to this need corporate developers invented first the shopping plaza, and then its more sophisticated derivative the shopping mall. The principle behind these is a simple one — in the absence of an existing commercial main street the developer constructs a single retail block, surrounds it with ample parking, and leases the individual stores. The first small plazas were built in the 1930s; after the war large versions of these became the accepted form for retail development in North America. They were usually long, single-storey buildings paralleling the road, fronted by the parking space, though sometimes they were laid out in a U-pattern, or, particularly at intersections, in an L. The logic was that for customers in cars, and in the suburbs that meant most customers, the parking lot had to be easily accessible; once they had parked they could do all their shopping on corporate property (Figure 9.7). The resulting landscape of wide highways, low retail buildings, and barren asphalt spaces punctuated by tall signs advertising the plaza, is a distinctive feature of post-war surburban developments throughout North America. It is perhaps nowhere better developed than along some of the arterial roads of New Jersey, where one can only marvel at the miles and miles of plazas and apparent complete lack of any centres.

The possibility of providing a more distinctly demarcated pedestrian area in a plaza and then enclosing it from the elements was first identified in the early 1950s in the harsh winter areas of the northern United States. In 1953 a climate-controlled shopping mall was opened in Omaha City in Nebraska, and malls in Minneapolis and Detroit soon followed. These were the corporate and North American contemporaries of European pedestrian precincts, and they proved to be equally successful. Their

Figure 9.7: Corporate plazas and malls to service the corporate suburbs: a 1970s L-plan plaza in Toronto and a fragment of Fantasyland in the West Edmonton Mall, probably the largest enclosed shopping mall in the world with over 800 stores, an ice rink, wave pool, miniature golf course and fun-fair. These are little corporate states, private indoor worlds that have replaced the public streets of previous centuries

widespread adoption was, however, quite slow, perhaps because of the high costs of construction. Nevertheless by 1980 every new North American suburb and many downtown redevelopments included enclosed shopping malls, and they were rapidly becoming an accepted part of new European landscapes.

Superficially shopping malls look like covered streets, an illusion which developers do little to shatter. In fact there are major differences between the two. Streets are public spaces; malls are wholly designed and under complete corporate control — they are great money-making machines exactly designed with profit in mind. On the basis of a mixture of common sense, trial and error, experience and behavioural research, mall designers have attempted to eliminate almost all unprofitable elements. Maximum parking distances from the mall entrances should be less than 400 feet, exterior walls should be blank but on the inside each store should have its own façade, the entrances should be well marked on the outside but on the inside should be as inconspicuous as is possible under fire escape regulations, a multitude of mirrors remind us of our shabbiness, and corridors should have sight lines of no more than 200 feet and be no more than 20 feet wide. Lighting and sound levels are controlled to keep people alert, benches and other furnishings are placed in the centre of the walkways so that shoppers are forced close to the always open store entrances, and the seats should not be too comfortable (flat wooden surfaces are common) so that shoppers do not waste good buying time sitting down.

In the latest phase of mall design the usual approach is to create a variety of small spaces and sightlines, with detailed store façades and textures, and to reproduce whole outside environments. First plants, then fountains, streams, statues and trees were moved inside. In the super malls of the 1980s there are full-size ice rinks and fun fairs. The West Edmonton Mall, which claims to be the largest enclosed shopping mall in the world, has a skating rink, 'the original replica of the Palace of Versailles fountains' (miniaturised), a small enclosed theme park (called Fantasyland), aquariums and aviaries, a wave pool, and a nine-hole golf course (also miniaturised) based on the Pebble Beach course in California. It also has over 800 stores, including eleven department stores, a fantasy hotel (you can sleep in, or otherwise employ, a 1950s Ford), advertises in and attracts customers on package tours from cities thousands of miles away, and is still expanding. For people living on the Canadian prairies it is becoming a place for spending mid-winter vacations that ranks only slightly behind California and Florida.

Commercial strips — from casual chaos to television road

Nowhere is there clearer evidence of a free market corporate economy going about its work of competing for consumers' money than on commercial strips. Blazes of colours, dramatic signs exhorting us to buy, billboards, fantastically façaded buildings covered in company logos and names, are the direct revelations of corporate presence. We know these names and products and develop strange loyalties to them, buying this brand of gasoline rather than that one, this make of hamburger rather some other. They are strange loyalties because by most reasonable standards the products of different corporations are indistinguishable; it is only the marketing which varies. It is out of this marketing variety that the landscape of commercial strips is created.

The commercial strip is a particular form of the bright-lights areas of city centres that developed early in the century and had neon and electric advertising signs to catch the attention of pedestrians. Times Square and Piccadilly Circus are, of course, the primary examples. Beginning in the 1930s, and then more swiftly after the war, this sort of electrified commercial landscape was exported to the automobile realms of the suburbs. Since car drivers experience a phenomenon, sometimes known as 'whizz-by', whereby the details of landscape disappear as a result of travelling at high speeds, these suburban commercial strips had to have larger signs, brighter colours, more exotic styles and less detail than most of their downtown cousins. The resulting landscape looked all right from a car passing at 30 miles an hour, but in a photograph or from a pedestrian's point of view it was little short of a hideous confusion of ill-formed spaces, dirty surfaces, rusty poles, vehicles, and competing signs (Figure 9.8).

The appearance of strips has changed considerably over the last 40 years because of changes in sign technology, increased planning controls, and because independent businesses have been replaced by corporate franchises. In the 1950s strips were comprised chiefly of independent businesses which adopted idiosyncratic styles and fanciful names and signs. The signs were mostly neon, flashing electric, or dramatic eye-catching, whizz-

Figure 9.8: Commercial strips in Dallas and Montreal that give some indication of the shift from competitive chaos in the 1950s and 1960s to the orderly and modulated 'television roads' of the 1980s, in which a relatively small number of corporate outlets compete tastefully and repetitively; as overprocessed as an evening of network television (see p. 216)

bang boomerang or paraboloid shapes. The buildings were often what Robert Venturi calls 'decorated sheds', simple boxes encased in exotic façades which may or may not have any relation to what is sold within. In some dramatic cases buildings were sculpted into the shape of hot dogs, ducks or wedges of cheese. In all of this little attention was paid either to the details of landscaping or to the way everything looked together. The result was a vibrant and chaotic landscape.

In the mid-1960s strips began to acquire a new style. This was partly accounted for by the rapid rise in franchised outlets. By 1965 there were 1,200 companies in the United States granting franchises and some 350,000 franchised outlets, accounting for 30 per cent of all retail sales (Boorstin, 1973, p. 429). In a franchise system, such as Burger King or most gas stations, an individual store-keeper leases the right to use the marketing system and design package of a corporation. Since economic success now depends in no small measure on product identification, the designs for franchise outlets stress simple, recognisable forms and colours rather than idiosyncracy. Furthermore, by the 1960s flickering neon had become unfashionable and was being replaced by backlit signs of brightly coloured plastic. Increasing attention was paid to design detail and quality, and different styles were developed for different types of locations, for instance Sunoco and the other oil companies developed specific styles for residential, commercial and industrial settings.

The trend to neatness continued throughout the 1970s, paralleling the rising demand for stylishness in house design. Chains introduced more subdued designs, such as McDonald's standard restaurant with the shingled mansard roof, wood trim and landscaping of evergreens and flower beds (Figure 9.9). These changes were accompanied by increasingly restrictive municipal regulations for signs and site planning. The result has been that the most recent commercial strips consist of a series of well landscaped, carefully designed, standardised buildings to meet the major consumer needs of servicing the body and the automobile. It is all very trim compared with what was built in the 1950s, nicely finished and not unattractive. It is also lacking in vitality and individuality. Like corporate residential suburbs this is all packaged, but instead of the corporate presence retreating into the background here each corporation has its own distinctive heraldry of a sign or logo, plus company colours and built-forms, so that those who are unfailingly loyal to a particular corporate

Figure 9.9: Two versions of McDonald's. The restaurant style was changed in the late 1960s from a red and white tiled box with a single red and white arch, to this rustic brick and shingle with mansard roof, television road version. In the 1980s many outlets have been adapted to their context — this one in Washington, DC is a short distance from the White House and has a doorman

brand of gas, muffler, hamburger or pizza can find it no matter where they may be. 'In a country of highly varied climate and terrain,' writes Philip Langdon (1985, p. 76) about America, 'thousands of cities and towns now look as if they were put together with interchangeable parts.' The same thing can be said of Australia, Britain, Belgium, almost anywhere. The landscape of interchangeable parts reveals an economy based on multinational corporatisation.

Corporatisation, planning and architecture

Theo Crosby, in his book *How to play the environment game* (1973), offers an account of modern urban development as a complex game between the great development corporations on the one side, and planners representing the state on the other, with the corporations usually winning. Since the 1940s the aggrandisement of development corporations and planning departments has certainly been parallel, almost as though they have grown in stature as a result of their contest.

The relationships between corporations and planning departments differ from nation to nation according to the details of planning legislation, and from municipality to municipality depending upon local politics and circumstances. Nevertheless, a common view is that planning acts as a constraint to the more outlandish activities of developers, such as William Zeckendorf's attempt to develop an elevated airport 200 feet above street level in the middle-west side of Manhattan between 24th and 71st Streets (Marriott, 1967, p. 232). It also ensures that minimum social provisions are made in any development. The narrow private interests of developers are thus made wider by planners representing the public interest.

An alternative view is to see planning as part of the development process. In much of the day-to-day administration of landscape change there is a considerable degree of co-operation and compromise between corporations and planners. Planning departments will grant development bonuses (usually so many extra feet of floor space) if, for example, a new skyscraper includes a publicly-accessible park or plaza, or if developers carry out certain road improvements.

A third and more jaundiced view is that planning is outdated and being sidestepped by corporations. Although official plans have the legislative authority of the nation or state behind them, they are mostly developed and applied at the municipal level. They are, furthermore, based on methods and models developed mostly in the 1920s, before automobiles became a wholly accepted fact of life, before television, before computers, before the International Style became popular, before megastructures, corporate tract developments, shopping malls, international fast food chains, jet travel, and above all before the omnipresence of multinational corporations. Thus municipal planning, based largely in outdated concepts, confronts international business employing the latest technologies and ideas. An immediate and common result of this imbalance is that if a municipality should reject a development proposal it considers shoddy, the developer can simply move to a less discriminating neighbouring municipality and build the project just over the municipal boundary. The corporations have time, money, flexibility and geography on their side, and the best that planners can hope to do is to moderate some of the worst effects of corporatisation.

So planning has to satisfy three roles simultaneously: it has to encourage the right contexts for development, protect the public interest against unscrupulous development, and struggle not to be outflanked by the financial and geographical flexibility of corporations. Crosby writes (1973, p. 239) that the planner 'spends his life on the defence, hopefully muting the worst excesses, administering a maze of complex and often meaningless regulations in the face of a continuous series of opponents, for whom any concession is a simple source of windfall profit.'

And what of architects in all of this? The corporate skyscrapers and new suburban headquarters are the masterpieces of modern architecture and the products of an apparently mutual admiration. It is in these more than any other buildings that the expectations of the Bauhaus and Le Corbusier have been realised, even as modernism has been the medium for expressing the progressive dreams of corporate executives. There is irony in this because modernist architecture was invented as a style for the masses, and many of the 1920s projects were socialist in inspiration and intent. When the Bauhaus architects migrated to the United States in the late 1930s they soon discovered that the new style worked equally well for corporate capitalist skyscrapers ('workers' housing stood on end and pitched up fifty stories', as Tom Wolfe puts it (1981, p. 4)). In the early 1950s Lever Brothers, Seagram, Union Carbide and a few others established the formula (that Geometric Steel and Glass Skyscraper = Symbol of Progress and Profitability), and then there was no looking back. The marriage between modernist architecture and corporate office had been achieved. It has endured.

In all the other branches of their activities corporations have paid little attention to modernism. Gas stations of some companies do still retain an element of the international styles they first used in the 1930s (and Mies van der Rohe did design a service station for Esso at Nun's Island in Montreal; 'Someone has to solve the problem,' he claimed, but his design proved too expensive to reproduce). Otherwise architectural principles are made subservient to perceived public demand — and that calls for almost everything that the modernists despised: brightly coloured signs, exotic façades, ornate brickwork, olde worlde styles and revivals, a kind of visual jamboree. It is these that are found along commercial strips and inside shopping malls and in corporate suburbs.

So the corporate attitude to modern architecture is to take it and leave it, as and when they choose. Great grey glass towers for the grey-suited executives and accountants who administer the companies; bright commercial revivals for retailing. The attitude of modernist architects has been mostly to take whatever commissions for office buildings come their way from corporations, to praise the results of these, and to sneer quietly at the gaudiness of the stores and outlets.

Commodification and the seductive corporate city

James Lorimer is a Canadian publisher who is less than enthusiastic about developers. 'The corporate city', he has written (1978, p. 79), 'is not so much a place for people to live and call their own as it is a machine rationally and effectively designed to make money.' He may be right, though you would never know it from the developers' blurbs promoting products and suburban houses and restaurants and stores. They are replete with references to community and neighbourhood, togetherness and contentment. In the corporate city each part of life is separated out and redesigned for greater profit while images of fun and romance take over from the grey realities of the world.

The process at work in much of this is one of 'commodification'. From the extreme business orientation of corporations everything is a potential commodity to be exploited, managed or otherwise manipulated in whatever ways will ensure the self-maintenance and profitability of the corporation. National symbols, local histories, fantasies and dreams, indeed anything nice from any time or place, can be imagineered to promote consumption. Thus we have McDonald's and clowns, housing developments marketed in terms of rural romanticism, package tours of the literary landscapes of England. Large areas of the town of Monterey in California have been commodified, chiefly around the fact that John Steinbeck wrote his novel *Cannery Row* about it (Figure 9.10). The 1984 tourist guide tells us that 'Vacationers follow the footsteps of departed ghosts: John Steinbeck's characters. Yes — on Cannery Row — the very place that Steinbeck struck life into the Row's wonderful figures ... the rotting buildings were remodelled into stores and shops and former brothels became restaurants.' Cannery Row has been turned into a strange parody of itself, with international European hotels, fish restaurant architecture covered in nets and navigation lamps, a 'Steinbeck's Lady' boutique, and luxury

Figure 9.10: Commodification of one of the old canneries on Cannery Row in Monterey, California, a street made famous by John Steinbeck's novel of the same name. Note both the Steinbeck's Lady Boutique and the standard fish restaurant decoration of thick ropes, jolly sailor and weathered planks

condominiums mimicking the building forms of the old canneries. None of this is unpleasant, the new buildings are interesting and carefully integrated with the old styles. Yet commodification has blurred the boundaries between fiction and reality, location and imagination, history and invention, until they are no longer distinguishable.

The people who conceive and promote such environments are not necessarily aiming to exploit others ruthlessly; like many of us they probably shop in the corporate centres, live in corporate suburbs, take their children to Disneyworld or themselves to Club Med for their holidays, and they know that the lifestyles and landscapes created by corporatisation are mostly pleasant and convenient. The fact is that corporatisation is extremely seductive, and its products are specific, tangible and conveniently available everywhere. Protests against it, while undeniably valid and important, seem to be mostly abstract or ideological complaints about loss of choice or the fact that commodification dilutes meaning; they seem to be arguing that people should not enjoy themselves and have comfortable homes. Perhaps this is why, in spite of the protests and expressions of concern, in the last 50 years corporations have been so enormously successful in rearranging city skylines, building suburbs and shopping centres, making theme parks and tourist resorts, and filling streets with their internationally recognised symbols and products. Corporatisation of cities is not yet complete — older fragments of cities remain to remind us that there can be different ways to organise cities and to make urban landscapes. But the process has such momentum and such seductive appeal that it is becoming harder and harder to imagine a future world that will not consist mostly of countless versions of commodified Cannery Row.

10

Modernist and Late-Modernist Architecture: 1945-

In the 30 years following World War II modernist architecture had its heyday. Almost everywhere, for corporate office buildings and new industrial buildings, for pedestrian precincts and reconstructed centres of blitzed European cities, for new towns and urban renewal projects, for hospitals and new universities, the rectangular Bauhaus/Le Corbusier style or some derivative of it came into favour. The last vestiges of classicism and pared down georgian did cling on for a while, chiefly in local shopping centres, and suburban houses followed their own semivernacular fashions, but the overwhelming trend was to a modernist style marked by plain surfaces and clearly defined shapes. This was a period of unusually active city building, and the fact that modernism was in fashion meant that it had a considerable effect on urban landscapes.

In the 1960s it really did seem as if modernism might last for half a millennium — there had been the gothic and renaissance styles, now there would be a great era dominated by a machine age architecture of concrete, steel, glass and plastic. This hope soon turned out to be false. In part this has been because 'postmodern' styles, filled with historical and regional allusions and remarkable for their applied decorations, and therefore antithetical to modernism, have become increasingly acceptable. At the same time there has been a rapid diversification and reworking of modernist forms in what seems like a mannerist attempt to push their special qualities to the limits or into complex arrangements. Modernist buildings continue to be constructed, so it is clearly inappropriate to call the movement dead, but given this evidence of diversification and post-modernism it seems justifiable to describe the period from about 1970 up to the present as one of Late-Modernism, that is, as its declining phase.

Mies van der Rohe and the skin and bones style

The International Style was neither widely nor swiftly adopted in the 1930s, in spite of all the enthusiasm and exhibitions. By 1939 there probably were still only a few hundred modernist buildings worldwide, a lack of success which was probably due partly to the decline of building activity in the Depression and partly to the preference of fascist and communist governments of the late-1930s for ideological variants of neo-classicism with lots of columns, peristyles and heroic sculptures. Most city streets and the buildings lining them still had an edwardian, victorian or older character. This began to change almost as soon as World War II was over.

When the Bauhaus was disbanded by the Nazis most of marxist members went to the Soviet Union and its more liberal members — notably Walter Gropius and Mies van der Rohe — found their way in due course to America. There, in the 1940s and 1950s, they reworked modernist architecture with its plain surfaces and straightforward rectangular shapes, they made it popularly acceptable and they adapted it successfully to corporate skyscrapers, apartments and university buildings. It was then exported back to Europe, where Le Corbusier had in fact been developing something similar in concrete rather than metal and glass, then to Asia, Latin America and Australasia. In one form or another modernism has come to dominate the appearance of the late twentieth-century city.

Mies van der Rohe was appointed to a position at the Illinois Institute of Technology at Chicago in 1938 and was given responsibility for redesigning the campus there. His plans (he drew up several along essentially the same lines) were simplicity itself. All the buildings were arranged in lines and at right angles, and their basic form was that of a carefully proportioned cube expressing the structure of steel columns and beams as perfectly as possible (Figure 10.1). The slogan Mies invented to express his design philosophy was 'Less is More'. The less part of this was clearly evident here, where everything was reduced to its simplest rational elements and all other considerations were subordinated to a rectangular geometry. Even the campus chapel was hidden in a cube, a fact which has caused the architectural critic Charles Jencks (1977, pp. 16-17) to claim that the most religious building on the campus is in fact the heating plant because it alone has something pointing heavenward — its chimney.

Figure 10.1: The rectangular buildings and lamp-posts on the rectangularly planned campus of the Illinois Institute of Technology in south Chicago. Designed by Mies van der Rohe and built at various dates between 1939 and 1956, these were a major source of inspiration for post-war modernism in America

When hoisted into the sky these sorts of angular geometric styles were clearly well suited to the steel frame construction of skyscraper offices and the existing grid patterns of streets in American cities. The first major post-war office building was Lever House, designed by Gordon Bunshaft for the firm Skidmore, Owings and Merrill and built in 1951 on Park Avenue in Manhattan. Not very high by subsequent standards, this is a slab of two-tone green glass rising above a podium that is itself raised on stilts one level above the street to create a semi-public plaza. Two things are especially significant about Lever House; first, it established the basic modernist style for corporate skyscrapers for the next three decades; and second it invented a way of building in the context of New York City zoning restrictions for height that did not result in a wedding cake skyline but left part of the available space at the lower levels unused so that the tower could rise sheer from a public plaza to its squared-off roof line. Both the plaza and the built-form were to be widely copied.

In the annals of modern architectural history the second great stride forward in skyscraper design occurred just across Park Avenue from Lever House about five years later. This was the Seagram Building designed by none other than Mies van der Rohe (in co-operation with Philip Johnson). Here Mies directed his fascination for less-is-more, geometrical perfection 40 storeys skywards. In architectural books the resulting tower is usually described as bronze and shown standing in splendid isolation, as it did when it was built. It is, however, more accurately described as black, and from street level is now almost hidden on three sides and shorter than the adjacent buildings. For the architecturally uninformed strolling along Park Avenue it no doubt appears as just one skyscraper among countless others; for those who choose to look carefully the Seagram Building has precise mathematical proportions and a rigorous order in all its parts, including constrained blind openings (they can only be closed, completely open or half open so that a messy exterior appearance is avoided). Order and elegant simplicity are paramount. The black/bronze surface and exactly repeated forms leave little to be contemplated except pure shape. It is a matter of opinion whether this is attractive or not, and the opinion of architects is by no means unanimous; nevertheless the overwhelming majority of them do recognise the significance of the Seagram Building — photographs of it appear in every discussion of modern architecture.

In the economic growth and city centre redevelopment that took place during the late 1950s and throughout the 1960s Lever House and the Seagram Building served as prototypes for the design of many office towers. This was 'skin-and-bones' architecture, in which the bones of the steel beams are expressed in the overall shape and in the grid of window and floors, and the skin is an almost featureless curtain wall of glass apparently hanging in front of the skeleton. It seemed simple enough and many architects tried to copy it; Mies himself reproduced the Seagram Building in slightly different versions in Pittsburgh, Chicago and Toronto, and it was widely plagiarised in various colours and shapes by other architects. In spite of its apparent simplicity the skin and bones copies rarely look as elegant as the originals (Figure 10.2). Indeed some of them are so devoid of interesting qualities they scarcely generate any reaction at all. 'Thinking, the other afternoon,' wrote John Updike in the *New Yorker* in 1962 (p. 107),

Figure 10.2: The curtain walls of skin and bones skyscrapers on Park Avenue in New York of which John Updike wrote that 'perhaps the first thing to say about the new architectural mode is that it leaves one with little to say. It glossily sheds human comment.'

that we ought to welcome the future to our city, we strolled over to say hello to the massive, glinting architectural new-comers that have suddenly filled ten or so blocks [of Park Avenue] — and discovered that once 'Hello' was pronounced, the conversation threatened to end. For perhaps the first thing to say about the new architectural mode is that it leaves one with little to say. It glossily sheds human comment.

And he departed 'glassy-eyed from contemplation of these buildings made entirely of windows.'

Le Corbusier and the concrete cage

Architects, it seems, are professionally inclined to mimic the work of their masters, and Le Corbusier has evoked possibly even more admiration and copying than Mies van der Rohe. His buildings were, however, more varied than those of Mies. The post-1945 Le Corbusier style took several forms but for the sake of clarity these can be grouped into two types — one angular and the other curved, or in more formal terms one cubist and the other expressionist. Both involved sculpted forms and textured surfaces of poured concrete, a material available for half a century which seemed to realise its potential for the first time in Le Corbusier's designs.

The prototype of the angular concrete structure is an apartment designed by Le Corbusier and built near Marseilles between 1947 and 1952 — L'Unité d'Habitation (Figure 10.3).

Figure 10.3: Le Corbusier's Unité d'Habitation in Marseilles, 1947-52, and his chapel at Ronchamp, 1950-5, are the prototypes for hundreds of concrete cages and poured concrete buildings everywhere

Source: Hitchcock, 1958, Figs. 166, 167

Significant partly because it was an ambitious attempt to combine apartments, community services, day care centres and shops in a single building, it is also visually striking because of its large-scale sculptural qualities. The façade is not a smooth surface, nor a simple rectangular grid, but consists of different rectangles as though superimposed, with some openings larger than others, and coloured recesses that accentuate the play of light and shade. A somewhat different angular character was given by Le Corbusier to La Tourette, a convent built in the late-1950s. With its patterns of intersecting beams and concrete panels this could be a huge cubist sculpture. An underlying order was given to all these mixed shapes and complex forms by the use of a system of proportions Le Corbusier devised in the 1940s which was based on human dimensions and which he called the 'modulor'.

The most famous of Le Corbusier's expressionist concrete structures is a church at Ronchamp in France, Notre Dame du Haut, built between 1950 and 1955. With its complex curved walls pierced by openings of irregular sizes, and a massive pointed roof which can be seen as representing hands held in prayer, the prow of a ship, a sail, a three-cornered hat, or many other things, this building is filled with symbolic suggestions. It inspired a number of other expressionist concrete buildings, such as Eero Saarinen's bird-like TWA terminal in New York. It also established with great clarity the distinction between the concrete/sculptural and the steel and glass/rationalist forms of modernism because Ronchamp could scarcely be further removed in form from Mies van der Rohe's shoe-box chapel at the Illinois Institute of Technology.

Le Corbusier's concrete structures, both angular and expressionist, seem to have been an inspiration for poured concrete architecture everywhere during the next decades. Boston City Hall, much of the South Bank development in London, Scarborough College in Toronto, all clearly owe a great debt to them. Somewhat less obviously derivative are the apartments and towers made of concrete sections and panels which try to achieve some surface texture and sculpting by means of balconies, recessed windows or patterns of columns beams (Figure 10.4). In many office buildings concrete has been used in precast sections, or repetitive forms poured in place, and most sculptural qualities have been pared away until only a gigantic concrete cage with recessed windows remains. In this 'concrete cage' style a light-coloured concrete frame either takes precedence over or is of equivalent visual prominence as the sculptural effects or the complete form. The cage itself may be given a horizontal or a vertical emphasis, but in either case is usually a simple grid in which the influence of Le Corbusier designs is echoed faintly in the concrete shapes. Whatever its design source the concrete cage has become one of the most common modernist styles, and large sections of urban landscapes, such as the downtown area of Houston and the office district of Canberra in Australia, seem to consist of almost nothing else.

Figure 10.4: A variety of concrete cages, close-up in Buffalo (the people between the two buildings give a sense of the monumental scale), and composing much of the downtown mass of Houston

Modernist architecture, then, has two primary styles: there are the smooth, elegant, skin and bones, less-is-more structures in the manner of Mies van der Rohe; and there are the textured, moulded, concrete-cage buildings probably derived from the designs of Le Corbusier. This distinction is not always sustained so neatly in landscapes. There have been other influences at work, other architects such as Louis Kahn and Alvar Aalto have developed different types of influential designs, and style features have often been mixed in single buildings in order to meet the wishes of corporate clients or to express originality. In spite of attempts to create distinctiveness it was soon apparent that modernist architecture does not permit many significant variations, especially in skyscraper offices and suburban industrial boxes. The dominant lines can be made vertical or horizontal or into a grid; the colours of the glass and metal can be changed; the proportions of the windows can be varied; the shape of the building can be altered to fit the site. All of these have been tried, tried again, combined and reworked in an effort to make strong identities for particular buildings, but the sheer bulk of most new offices and their lack of detailed decoration seem to render most such efforts at originality inconsequential. The result is that most modernist buildings are almost indistinguishable one from another. They merge both in perception and in memory into a confusion of pale tones, sharp angles and cereal box forms, standing on end downtown and lying flat in the suburbs. This anonymous aesthetic does not sit well in a capitalist and competitive economy, and about 1970 corporate owners apparently decided that enough anonymity was enough and began to adorn the top storeys of most new offices and factories with brightly coloured signs and logos. They presumably wanted to be recognised.

No-frills modernism and new brutalism

There is a third type of modernist architecture, probably more widespread than either skin-and-bones or concrete cages, though it is less eye-catching, and has far poorer credentials and a more obscure line of descent. It is not clearly influenced by the designs of the hero architects. Rather it is the work of local architects and engineers adapting modernist styles to immediate needs and limited budgets. The result is what might be described as 'no frills modernism'. It is the mostly anonymous vernacular architecture of the new industrial state (Figure 10.5).

Industrial state vernacular has the rectangular forms and unadorned surfaces of all modernist architecture, thereafter the similarities are few. There is little elegance or concern for proportion and perfection of finish; construction materials are ordinary and used in combinations — bricks with concrete blocks, laminated panels, prefabricated sections, aggregates, squarish windows of standardised dimensions. The roofs of

Figure 10.5: No-frills modernism or industrial state vernacular. There are countless structures which have a plain, angular look about them, but lack the style, grace and proportion of the best modernist buildings. The group of houses and apartments are in west London. The street scene is in suburban Toronto; no-frills modernist buildings include the bus shelter on the left, the gas station and the apartments in the background

houses, like those in English housing estates, may have a low pitch, otherwise they are flat and often have rooftop machinery showing. Stores in plazas and buildings in industrial subdivisions may have a partly decorated façade, perhaps of stucco or stone cladding, which they present to the public, or at least to the street; around the side or the back there are concrete block walls, mass-produced windows units and windowless metal doors.

The landscape consequences of all of this are remarkable only in their ordinariness. It is hard to get excited by, or indeed even to notice, these buildings. They include small apartments, municipal garages, schools, multi-storey car parks, factories and warehouses, lesser government offices, the rows of shops in 1950s and 1960s plazas, bus shelters and gas stations, in fact all the poor relatives in modern building. At their worst they are almost devoid of positive aesthetic qualities — little more than boxes with squarish holes for windows and oblong ones for doors, architecture reduced to degree zero. More commonly they have some element, a portico, or a cluster of windows, perhaps a carefully designed sign, which make a token gesture towards the clean-cut elegance of their more expensive relatives.

The predecessors of this no frills architecture lie perhaps in some of the less spectacular achievements of the Bauhaus, in projects for workers' housing in Germany and the Netherlands, in utilitarian industrial buildings, and, in America, in standardised gas station designs of the 1930s. In Europe war-time shortages introduced further architectural simplicities, such as Quonset Huts, brick-box air-raid shelters, and prefabricated houses made of asbestos panels. All of these prepared people for plain and homely designs made of mass-produced, standardised and readily available materials which may look drab but which are far cheaper than the custom-made components of architect-designed structures.

In the 1950s a small group of European architects, including Peter and Alison Smithson in England and Aldo van Eyck in Holland, did try to formalise this industrial state vernacular. By self-consciously using industrial materials, such as metal stairway units, by revealing evidence of skilled workmanship in the details of windows and doors, by exposing building materials which would normally have been hidden, by leaving heating ducts and hot water pipes in full view, and by being brutally honest in expressing the functional realities of a building, they hoped to 'humanise' modern architecture by making it less intellectual.

The austere style which resulted came to be known as 'New Brutalism', a catchy but confusing name because the attempt was to create a humane welfare state architecture, and because it has an etymological proximity with *beton brut* — raw concrete. There is grist for the academic mill in this, and depending on who you read Brutalism can refer to 'a tough-minded reforming movement with the framework of modern architectural thought' (Banham, 1963, p. 64), 'a taste for the intimidating, the gratuitously hostile' (Drexler, 1979, p. 11), a glass and brick vernacular style which makes a 'direct reference to the socio-anthropological roots of popular culture' (Frampton, 1980, pp. 263, 266), or poured concrete buildings with imprinted board forms. Regardless of such debates it is obvious that in practice Brutalism offers a neat aesthetic justification for cheap no-frills modernism (or perhaps just cheap-looking modernism) in which things like pipes and air circulation ducts can be left exposed and the cost of installing ceilings can be avoided. I suspect that the users of Brutalist buildings, most of which seem to be institutions of some sort, assume that their industrial quality environment is just another example of government penny-pinching and would be amazed to learn that it has been self-consciously designed to look like this.

Expressionism

By the end of the 1950s the worst hardships of the aftermath of the war had been transcended, cities began to expand upward and outward, new towns and new shopping centres had to be built, schools, highways and offices were needed. The architects who responded to this demand had at their disposal the skin-and-bones geometric style of Mies van der Rohe, the cubist sculptural style of Le Corbusier, and the unassuming versions of Brutalism or industrial state vernacular. What happened after about 1955 was that each of these proliferated, though undergoing changes and hybridisation, until about 1970. In that period it would have been virtually unthinkable for a building, except perhaps something on a commercial strip or in Disneyworld, to be designed in any style other than one of these variants of modernism.

There was one acceptable alternative to unornamented rectangularity, and that was self-conscious modernist expressionism in which the architect attempted to express some idea or symbol in an imaginative way. This had been a distinctive movement within modernism ever since the early 1920s when Eric Mendelsohn had designed the Einstein Tower in Potsdam, a six-storey observatory of poured concrete that looked more like a gigantic plant than a scientific laboratory. In the 1950s and 1960s, perhaps inspired by Le Corbusier's expressionist chapel at Ronchamp and perhaps because there was more money available for imaginative projects, expressionism became quite fashionable for institutional buildings, city halls and churches. These almost always have either the exaggerated sculptural qualities of poured concrete or the sleek finishes of modernism. It is the forms which are so dramatic and unusual. The Guggenheim Art Gallery by Frank Lloyd Wright is made of spirals like an enormous snail shell, the two curving towers of Toronto City Hall seem to embrace a flying saucer which houses the council chamber, and the cluster of huge mussel shell forms that comprise the Sydney Opera House marvellously echo the sails of ships in the harbour. The chapel for the Center of World Thanksgiving in Dallas has a spiral form reminiscent of the path to heaven depicted in old drawings to illustrate *A Pilgrim's Progress* (Figure 10.6). Expressionist buildings like these are, however, still quite rare. They attract a great deal of attention precisely because they are so unusual.

Late-modernist architecture

After 1970 the single-minded commitment to the main forms of modernism began to weaken. Post-modernism and heritage preservation became increasingly important, and modernism itself was elaborated into a confusing number of variations and sub-styles. The conventional boxy modernist looks of the 1950s and early 1960s did not cease to be built, but were now joined by reflecting glass cubes, blank concrete boxes, self-conscious displays of structural ingenuity, and complex arrangements of shapes. This diversification of modernist styles has proceeded so rapidly over the last 15 years that it seems legitimate to regard it as an indication that modernism is in a declining or late phase in which most of its qualities are being increasingly exaggerated — buildings are becoming either less ornamented than ever, or increasingly conspicuous structural demonstrations with more complex forms.

Figure 10.6: Expressionism: the Spiral Chapel of the Center for World Thanksgiving in Dallas. Expressionist buildings are intended to reveal some idea or conviction and are more popular with governments than corporations. In this case the spiral is a traditional Christian symbol of the difficult pathway to heaven; it contrasts dramatically with the rigidly modernist landscaping and grey wall of standardised panels of the skyscraper behind it

There are difficulties in attempting to classify or to comment on late-modern architectural styles, partly because these are still developing and have not necessarily begun to present themselves clearly, and partly because the built-forms have begun to outrun any available architectural language to describe them. They could be interminably elaborated and subdivided. From the perspective of their role in recent landscape changes I prefer to identify just three distinctive late-modernist forms — Blank Boxes, Glass Boxes and the Engineering Style, and two widespread modifying qualities to these which I call 'sleekness' and 'forced originality'.

Blank boxes

With sophisticated air circulation and lighting technologies (and by ignoring problems of Legionnaire's disease and internal smog) it has long been possible to build windowless structures. This fact is pushed to its logical conclusion in blank-box buildings. A perfect blank box consists of a flat roof on four walls punctuated only by doorways. Such structures are not uncommon in industrial subdivisions because they make excellent warehouses. Elements of this approach are, however, frequently used on more prominent buildings, such as the new wing of the National Gallery in Washington, DC, though rarely with a complete exclusion of windows. The result is often a minimalist one emphasising severe shapes, blocks and blank walls (Figure 10.7). A particular variant of the blank box is a 'fortress style', with great bastions and beams around the main entrance, often coupled with sheer blank walls, apparently in order to stress the massiveness and potential impregnability of the building. I am uncertain why.

Glass boxes

The possibilities of pure glass architecture were conceived in the early 1920s by Mies van der Rohe and others, but their realisation had to await technological developments in mounting the panes and in reducing solar gain — the heat build up that turns glass buildings into hothouses. Both of these problems were overcome in the 1960s and early 1970s (though not without some mishaps, such as the panes of glass on the John Hancock Tower in Boston which popped out and spiralled down to the streets below, no doubt to the considerable surprise of pedestrians). The new mountings allowed continuous glass surfaces, and solar gain was reduced by implanting a very fine layer of reflective material within the panes. Since glass is a cheap way to put walls on metal frames the result of these technological developments has been a profusion of mirror glass buildings. Many of these, Drexler (1977) suggests, 'are so refined as to communicate almost nothing' (p. 12); they reflect the sky and other buildings and they offer interesting plays of light tinted by the tone of the glass; only at night, when the interior lighting is on, do they become transparent, a figure-ground reversal on a grand scale (Figure 10.8).

Figure 10.7: Blank boxes: the World Trade Center in Dallas (almost every city now has a world trade centre); the dramatically malevolent black building in a Darth Vader style is behind New York City Hall and houses the Family Court; the fortress entrance to the new wing of the National Gallery in Washington, DC

Glass box forms have gone through a rapid evolution. The first ones were usually glass cubes, flat surfaces on all sides. Then ways of curving surfaces were found, and both convex and con-cave shapes appeared. More recently the surfaces and corners have been bevelled, chamfered and cantilevered, and the forms have been elongated, sculpted and grouped so that many of the newest mirror glass buildings look like gigantic clusters of crystals catching the sun and the changing light. These can be visually spectacular, and crystal clusters such as the Renaissance Center in Detroit and the Westin Hotel in Los Angeles have quickly

Figure 10.8: A glass box and a crystal cluster: a recent addition to the Exhibition Building in Melbourne, and the Renaissance Center in Detroit with a hotel tower surrounded by four office towers

assumed roles as major landmarks in the skylines of their respective cities.

The engineering style

This seems to be an extension of Brutalism. It is the style in which the structure, materials or building components are celebrated, and form becomes secondary. Or perhaps, as Banham (1984) argues in *The architecture of the well-tempered environment*, this is an indication of the pre-eminent role of what he calls 'environmental technologies', those concerned with heating, lighting and air circulation and conditioning, in modern building. In the Pompidou Centre in Paris, for example, these have virtually eliminated modernist forms. There are many ways of responding to some supposed technological imperative, but the commonest is perhaps some sort of 'hi-tech' approach in which apparently industrial and functional materials are employed stylishly — with metal railings, exposed elevator workings and ducts and pipes, metal siding, bottle glass windows, stairwells stuck on the outside, and so on (Figure 10.9). Another approach is to make a building into a structural demonstration, for instance by employing an exoskeleton, or dramatic cantilevers, or some structural system which allows future modular units to be clipped on. A further possibility, used especially in the buildings belonging to the manufacturers of building products, is to stress materials; hence, a skylight company adorns its buildings with skylights, and steel companies in Pittsburgh and Toronto have buildings with a steel cladding which is allowed to rust (the surface rust provides a protective layer against deep rust penetration). Finally, there has been a minor revival of Crystal Palace greenhouse styles, using prefabricated sections to create winter-gardens, arcades and atriums. In all of these variants of the engineering style a building makes a technological/industrial statement rather than impressing by its shape or surface.

These three styles of late-modernism have been modified by two qualities — 'sleekness' and 'forced originality'. Sleekness refers to an exaggerated concern with finish, whether of form or surface. It is apparent in smooth surfaces of glass or gleaming white panels split by ribbon windows, in precise lines, acute angles, stainless steel details, space-age streamlining and the same sort of concern

Figure 10.9: The engineering style: Xerox Research Building, Mississauga, Ontario, and the Wintergarden Pavilion, Niagara Falls, New York. The engineering style involves a self-conscious display of structure or material. Buildings may be distinguished by cantilevers, greenhouse cages, exposed ducting and chimneys, industrial quality materials like thick glass blocks, awkwardly combined forms and/or colours that look like primer paints

with postured elegance that is found, for example, in high fashion. At its most extreme sleekness results in a high stylishness which brings a combination of precise form and finish close to perfection (Figure 10.10). More generally the desire for sleekness seems to have meant that the no-frills materials and industrial state vernacular have become less and less acceptable, and even the cheapest new buildings, especially when compared with their equivalents of the 1950s and 1960s, now seem to aspire to an

Figure 10.10: The sleek style of late-modernism: Consolidated Bathurst Building, Mississauga, Ontario. This sharp white wedge with a horizontal window strip has the essential features of sleekness — plain white surfaces, ribbon window, accentuated and elegant forms

architectural *haute couture* rather than being content with work clothes.

'Forced originality' is the term used by the novelist Isaac Bashevis Singer to describe a building shaped like half a boiled egg in Albany in New York State (Kennedy, 1983, p. 77). It is a phrase which exactly describes a widespread tendency in recent architecture — the self-conscious striving to find new and different shapes or combinations of materials for no apparent reason other than to be different (Figure 10.11). The late-modern breakaway from the rectangular constraints of early modernism has involved the use of obscure geometric forms, the incorporation of curved corners into buildings, and the creation of hybrids with bizarre and complex forms. The most recent generation of late-modern architecture includes buildings which combine several levels and towers in a single structure, which have irregular ziggurat shapes, which juxtapose reflecting glass sections and blank concrete walls, or mix greenhouse atriums with structural displays, cantilevers, sleek finishes and hi-tech details. Whether it

Figure 10.11: Forced originality: the impossibly poised City Hall in Dallas, perhaps built upside down, is a mixture of expressionist, engineering and blank-box styles; a vaguely pueblo style apartment complex overlooking the River Thames in west London

comes from the architect or the demands of the client, forced originality is the expression of a desire to make each building visually distinctive, a desire which is taking late-modern architecture ever more rapidly away from the conformist approaches of the 1950s and is contributing a self-consciously stylish and pointless eclecticism to the urban scenes of the late twentieth century.

11

Post-Modernism in Planning and Architecture: 1970-

Every change in environment is criticised by someone. The bigger the changes the greater are the criticisms, so modernisation and modernism have come in for resounding condemnations. In 1955 Ian Nairn and Gordon Cullen surveyed the state of the modern English landscape for *The Architectural Review* and expressed their reactions in an article simply called 'Outrage'. They drove from Southampton to Carlisle, looked at a mess of signs, arterial roads, fences, wires, suburban bungalows, institutions for the insane, sanitation plants, and 'the wreckage of wars and War Departments', and they despaired. Their account was, they declared, less of a warning than a prophecy of doom; if development continued in this fashion by 2000 Britain would consist of isolated monuments in deserts of wire, concrete roads and bungalows.

This feeling of outrage about modernisation was widespread. Peter Blake expressed it about American landscapes a few years later in a book he called *God's own junkyard*, filled with photos of meretricious or downright ugly scenes. Christopher Tunnard and Boris Pushkarev (1963) were more specific but no more positive when they noted the disparity between private goods and public environments: 'The expertly engineered automobiles', they wrote (p. 3), 'are seen against a background of ramshackle slums ... the spreading suburbs seem to have no centers of life or individual distinction.' Then Jane Jacobs (1961, p. 4) took a resounding swipe at urban renewal: 'Low-income projects are worse than the slums they replace, middle-income housing is a marvel of dullness, luxury housing vulgar ... This is not the rebuilding of cities, this is the sacking of cities.' So by the end of the 1960s a significant body of opinion had emerged which held, in essence,

211

that modern landscapes didn't look good and probably didn't work well either.

In the 1970s these criticisms rose to a crescendo. Robert Goodman argued that modernist architecture was repressive, an expression of the arrogant authoritarianism of planners, politicians and corporate leaders. Brent Brolin wrote a book called simply *The failure of modern architecture* (1976). Ada Louise Huxtable, the architecture critic for the *New York Times*, claimed (1972, p. 11) that 'What is being designed ... is instant blight. We are building blight for the next hundred years. The environment is being sealed into instant sterility and its social problems are being compounded by ... substandard design.' Peter Blake must have agreed because in a second book, *Forms follows fiasco* (1977), he resumed his attack on modern environments by systematically ridiculing each sacred cow of modernism (he called them 'fantasies') — functionalism, the open plan interior, purity of form, lack of ornament, technologies of mass production, skyscrapers, zoning, mobility and the ideal city. As a start towards dealing with all this ugliness and failure Blake recommended (p. 157) a moratorium on the 'dimwitted notion' of zoning, on the construction of high-rise buildings and new highways, and on the destruction of all old buildings.

By 1975 there was hardly anyone left who had a kind word to say about modernism. Even Philip Johnson, the architect who had helped import it into America in the 1930s and had been one of its greatest practitioners, declared that 'Modern architecture is a flop ... there is no question that our cities are uglier than they were fifty years ago' (cited in Blake, 1977, p. 10). Magnificent in the abstract and in isolation, modernism had in reality and *en masse* turned out to be repressive, ugly, sterile, antisocial and generally disliked. Since there was now so much of it around this presented a considerable problem. One quick and direct solution would be to blow it all up. Not very realistic perhaps, especially since there were still no clear alternatives, but most criticisms of modernism do include a suggestive set of photographs of the demolition of part of the Pruitt-Igoe housing project in St Louis (for example, Wolfe, 1981, p. 81 and Blake, 1977, p. 155; see also Coleman, 1985, p. 7, for an English equivalent). This modernist project of geometrically arranged apartment slabs had won a design award before it had been built in the 1950s; it proved to be virtually unlivable. In retrospect its demolition in 1972 seems like a symbolic act devised to announce the end of modernism.

The old is new again

Of course modernist methods of making landscapes have not dropped abruptly out of use. Especially in their late-modern forms they are still adhered to by many of those who developed them, and they find great favour with bureaucrats because there is apparently nothing easier to administer than a rationalistic concept enclosed in right angles. Nevertheless, about 1970 a deep change of heart seems to have passed through the design and planning professions. Whether this was just a response to all the criticisms that had been made to modernism is not clear; it seems more likely that it was encouraged by many factors, including the shift in social and political climate that had been brought about by popular grass-roots protests such as those against environmental destruction and the Vietnam war. Furthermore, a new generation of architects and planners wanted little to do with the stereotypical techniques of modernism which had been invented by their teachers, and which had an unwelcome habit of erasing the details and diversity of everything in landscapes that preceded them. In addition, the increased ease of international travel may have begun to make people more aware of the qualities of distinctiveness in different places, and increased affluence then given them the means to be able to demand some of these qualities in their own cities.

Whatever the reasons, in the early 1970s new approaches in architecture, now usually referred to as post-modernism, emerged more or less simultaneously with attempts to revitalise the old fabric of inner city areas, with a rise of interest in heritage preservation, and with new approaches to urban design and community planning. Post-modern architecture literally refers to what comes after modernism, but it is largely based on a self-conscious and selective revival of elements of older styles, and this is exactly what has been happening in revitalisation, preservation and urban design. The indications are that post-modernism is not just an architectural style, but an attitude which has infused almost every aspect of urban landscape making (Figure 11.1).

We are perhaps too close to these changes to form a clear conclusion about their implications and the exact character of the processes they involve. Because they seem to be undoing the previous 50 years of landscape making it is tempting to imagine that they are the beginning of a quiet revolution in how cities are made and maintained, and that repressive architecture and

Figure 11.1: The old is new again. A copy of a nineteenth-century farmhouse (with three-car garage), built in 1985, and a vaguely victorian business village, both near Toronto. In the 1980s old-style is fashionable

planning by great corporate or government bureaucracies is being replaced by more sensitive and varied alternatives. There is, however, an equally real possibility that these changes are no more than decorative or stylistic shifts in landscape making. This is, after all, an age of affluence in which utterly different life-styles can be affected almost at will, and in which quality of finish is often considered more important than underlying substance. It could be that these new, post-modernist ways of making diverse landscapes and decorated buildings are part of an imagineered disguise for continuing corporatisation. Such speculations cannot easily be confirmed or denied. What can be said with certainty, because the evidence is everywhere to be seen, is that the newest urban landscapes are not modernist but quaint, stylish and decorated. The old, it seems, is new again.

Commercial and residential gentrification

Writing in 1966 Robert Venturi, who was soon to become one of the front-runners of post-modernism in architecture, declared that 'it is perhaps from the everyday landscape, vulgar and disdained, that we can draw the complex and contradictory order that is valid for our [American] architecture' (p. 103). This was radical stuff; the everyday landscape he referred to was that of the suburb and commercial strip, not merely disdained but scathingly condemned by most architects. Modernism had never wholly succeeded in the landscape of retailing, except perhaps for a few gas stations and some publicly-financed and planned projects in Europe, such as the centres of Swedish new towns. Given the least opportunity storeowners have always announced their presence by whatever means are to hand — bright colours, lurid signs, fanciful names, exotic statutory and decorative façades. All of these were commonplace on commercial strips and in the tourist areas of city centres long before architects rediscovered their 'complex and contradictory order' and stylised it sometime in the 1970s. Howard Johnson's restaurants, for instance, had adopted a New England town hall-church style, with a spire and an orange roof, in the 1920s, and then persisted with it so that by the end of the 1950s there were several hundred of them across America.

Since commercial buildings involve an ephemeral architecture, one that wears out quickly and needs frequent replacement, they are good indicators of changing popular fashions. The fact that in the 1950s so many of them had been decorated more or less garishly was one indication of the public dislike of the purity of modernist styles, though the sharp angles and smooth vitrolite or tiled surfaces did owe something to modernism. In the late 1960s these garish, quasi-modernist styles began to undergo a profound change, first in coffee shops such as Denny's, and then in most fast-food restaurants (Langdon, 1985). The new fashion was one of wood beams, fireplaces, brick exteriors, wood framed windows, and landscaping with low-growing plants that would soften the appearance of the buildings without obscuring them. In due course most of the large franchise chains developed relatively subdued, almost domestic, façades for their outlets, invested in landscaping instead of leaving a building exposed in an asphalt desert, created different types of façades for different contexts (McDonald's has 16 stock façades currently in use), and even began to renovate old properties. More localised chains and expensive suburban restaurants recreated entire vernacular styles, eveything from fake châteaux to fake Colorado mine-shafts, depending on what seemed appropriate for the food served. Distinctiveness, albeit bogus distinctiveness, coupled with a sort of domestic cosiness and suburban trimness, had by the 1980s become popular. Not everyone has greeted this stylish tidiness with untrammeled enthusiasm. Kent MacDonald, writing in *Landscape* in 1985, describes it as 'Television Road', and finds it as overprocessed and denuded of vitality as an evening of network television.

While all this overprocessing was taking place in commercial strips, similar shifts in appearance were occurring in shopping plazas and with houses. In the 1950s these were mostly in the styles of no-frills modernism. Then sometime in the late 1960s small experiments in eclecticism were made, with every store in a plaza being given its own pseudo-vernacular façade — clapboard next to georgian next to tudor and so on. This sort of decorative ebullience was soon restrained, and by the late 1970s entire plazas were being built in a single quasi-historical style, or perhaps in sympathetically mixed styles so that they had something of the appearance of the main street of a small town (Figure 11.2). St Andrew's Village, a small shopping mall built near Toronto in the early 1980s, is described in its publicity brochure as having 'the character and charm of a century old village with ideas borrowed from the past to capture the spirit of the buying public's

Figure 11.2: From commercial eclecticism to thematic design. An eclectic plaza in Minneapolis, probably built about 1970, anticipating what was to be called post-modern architecture. This crude eclecticism has given way to lushly landscaped developments based on a single theme, like The Barnyard, Carmel, California, a plaza of pseudo-barns tended by janitors wearing coveralls and straw hats

"Back to Roots" movement — different elevations, staggered frontages ...' The Barnyard near Monterey in California is an upmarket plaza consisting of replicas of eight Californian barns-cum-stores selling Scottish woollens, pewter, needlepoint and the consumer goods for the internationally affluent, all grouped around a lushly landscaped courtyard tended by janitors wearing blue coveralls and straw hats.

St Andrew's Village and The Barnyard are entirely new, designed to create the 'ambiance' of old places. Ambiance, a word used to describe anywhere with environmental style and atmosphere, is a key concept in this sort of commercial design, and indeed in almost everything to do with post-modern land-scapes. Similar processes of ambiance creation have been at work in city centres, where old warehouses and markets, or entire streets of formerly run-down stores, have been reclaimed for bou-tiques and cafés, and given an appropriate street decor of inter-locking brick sidewalks, old-fashioned lights, bollards and signs (Figure 11.3). This often looks like historic preservation, but it is not that so much as a process in which old buildings, sanitised

Figure 11.3: Commercial gentrification. Newburyport in Massachusetts, an old whaling town that has been revitalised using many of the standard devices of gentrification — antique store signs, brick sidewalks, bollards, sandblasted brickwork

and adapted for the needs of the late twentieth century, are given a commercial appeal. To enhance this appeal, and to ensure commercial success, history sometimes has to be freely reworked. At Covent Garden in London the old fruit and flower market is now filled with boutiques and street theatre groups; in Georgetown in Washington old factories lining the Chesapeake and Ohio canal have been renovated into shops and condominium apartments. Examples of such 'adaptive reuse' can now be found in almost every city and town. Sometimes the developers take great liberties, and sometimes it is done with great attention to historical accuracy in the architectural details. Donald Appleyard maintained that in either case what happens is a form of 'disembowelment', as mixed use and working-class districts are replaced by trendy shops and the offices and apartments of upwardly-mobile professionals. This entire process of ambiance enhancement through adaptive disembowelment can be simply described as commercial gentrification.

The idea of gentrification is more usually used to refer to a process of residential improvement in which streets formerly occupied by relatively poor, working-class populations are upgraded and renovated by young, enthusiastic professionals who are searching for ambiance in the areas where they choose to live. This they have generally found in the architectural qualities and urban conveniences of areas such as Islington in London, Society Hill in Philadelphia or Cabbagetown in Toronto. Commercial gentrification is merely the business and retail version of this, employing many of the same stylistic devices and creating landscapes with a similar ambiance to that of residential gentrification.

The landscape evidence of gentrification is unmistakable: formerly unassuming victorian buildings, with peeling paint and a patina of having been lived in and used for a century, are subjected to sandblasting, cleaning, repainting and repointing; they are then fitted with new doors, brass letter-boxes, wrought iron fences and plants in the windows, so that they manage to look both old and brand new all at once (Figure 11.4). In a matter of only 15 years gentrification has spread like a wave across old residential districts of cities, its breaking edge marked by the vans of contractors and skips to carry away the rubble. It has not been without its social difficulties. As working-class populations have been replaced by middle-class residents the character of neighbourhoods has changed drastically. In Toronto, for example,

Figure 11.4: Houses during and after gentrification in Cabbagetown (now called Old Cabbagetown in gothic script on street signs), Toronto; porches are removed, windows are reglazed, brickwork is cleaned and repointed, woodwork repaired, the front yard is provided with a neat wrought-iron fence and statues of pets

streets that were formerly dominated by community life, with people sitting on front porches in summer evenings and children using front yards for play, have been privatised. Life has turned away from the public space of the street to the private space of the houses. The porches have often been removed in the process of renovation, the front yards which were open to the sidewalk have been fenced off and turned into small display areas, the streets have been given little signs which officially identify them as being in Cabbagetown, and the community now comes together just once or twice a year for newly organised street festivals. Like their post-modern commercial counterparts these gentrified residential streets are overprocessed, television landscapes which exude a pleasant air of make-believe.

Heritage planning and preservation

Under the futurist philosophies associated with modernism the preservation of things historical was reduced to the tokenism of fencing off places thought to be vested with political or historical importance, such as the birthplaces of presidents. History was rarely allowed to stand in the way of progress, though sometimes old buildings were saved by moving them lock, stock and barrel to a safe site.

In the late 1960s, perhaps because of the rate of destruction of attractive old buildings and their increasing scarcity, historic preservation took a forceful turn. In Britain in 1966 government-funded studies of Bath, Chester, Chichester and York identified both the need for urgent action to protect historic districts and the possible bases for policies to achieve this. The following year the Civic Amenities Act was passed, and this required all local planning authorities to designate conservation areas; these were not necessarily just to be preserved like oversized antiques, they were to be adapted to new demands and uses. In the United States the rise of concern with historic areas rather than individual monuments was contemporary with that in Britain. In 1931 there had been only two cities with such areas, by 1959, 20 cities had designated them, then in 1966 the Historic Preservation Act gave substance to the need to protect historically and architecturally valuable districts 'having special meaning for the community', and by 1975 over 200 cities had designated heritage areas (Tunnard and Pushkarev, 1963, p. 409; Appleyard, 1979, p. 46). Such legislation has now been duplicated in many countries, and urban heritage programmes have been adopted not only in Europe, Australia and North America but also in cities such as Zanzibar and Beijing.

In practice most heritage legislation does not have many teeth, and serves mainly to make it difficult for developers to demolish buildings that have been designated as historically significant. It does, however, constitute some legislative authority, and this, combined with a steady increase in public interest, effective lobbying by heritage groups, and the commercial success of many restored buildings and districts, has resulted in a remarkably sudden turn around of attitudes. Old buildings and districts have ceased to be considered simply as impediments to rational planning, and are regarded now as invaluable components of the post-modern landscapes of cities (Figure 11.5).

Figure 11.5: Historic restoration leaves a variety of marks on urban landscapes. The Monroe Block in Detroit waits for funds for the restoring to begin. The Old Town in Quebec City has been restored and reconstructed to conform to an ideal French Canadian past — these apparently old buildings never existed at the same time until now

When a real town is used as the set for a historical movie all the modern elements are removed or disguised, the road markings covered with straw and so on; this process is called 'aging'. Restoration involves an equivalent, though opposite and more permanent process of 'newing'. Exteriors are cleansed of grime and soot, weathered materials are replaced, parts that have been destroyed by the ravages of time may be reconstructed on the basis of historical records and meticulous research, and anachronous additions such as plastic signs are removed so that the buildings can be returned to an apparently pristine state. As though to reinforce this effect historic areas so often landscaped with antique-style street furniture, quaint old lamps, hitching posts, and horse troughs; if there are still cobblestones they are exposed, and street signs are often replaced by suitably historical ones with gothic script. The use of the district almost always changes. Historic districts are popular tourist venues, so some of the buildings are usually taken over by tourist-oriented restaurants and boutiques; others become museums and information centres, while some are turned into the offices of designers, architects, lawyers and similar professionals who seem to find heritage districts attractive and good for business. The overall result is a spanking new townscape paradoxically made of partially old structures.

Sometimes more than this is involved. In the Old Town in Quebec City, for example, a district restoration project involved both the renovation of existing old buildings and the reconstruction of French Canadian buildings which had been destroyed; some of these never existed at the same time, so a townscape has been re-created not as it once was, but as the restorers wished it might have been. This is not just historical idealism. The entire preservation process is wrapped up in ideologies and biases; it is not The Past which is restored or preserved, but someone's or some group's selective image of their preferred past. Lowenthal and Binney, 1981, argue (p. 17) that preservation of old structures is crucial to a sense of social and national identity, and indeed we find that almost every Dutch town has its preserved windmill, every Canadian city its pioneer village, and every English town its ancient monument. This concern with local and national identity through history has become especially pronounced in the multicultural cities of the 1980s where recent immigrants do not necessarily share the history of the founding fathers, and the historical depth and continuity of place has to be self-consciously established by heritage programmes.

In less than 20 years heritage preservation has had a wide-spread if fragmentary effect on the look of cities. Perhaps more important than this is the fact that it has helped to establish a popular appreciation for the qualities of old buildings and land-scapes in the face of futurism and the increasingly forced origin-alities of modernism. After 50 years on the fringe the past now serves equally with the present and future as a source for design ideas. There is a considerable irony involved in this because with the rapid changes induced by modernism the definition of what is old, and therefore worth making an effort to preserve, has come closer and closer to the present. Several of the famous early build-ings of modern architecture have already undergone their first restoration, for instance Sullivan and Adler's Guaranty Building in Buffalo, and the Chrysler Building in New York. The Bauhaus itself has, I believe, been turned into a museum of modernism and restored to its original state, while a proposal to demolish Lever House and replace it with a larger office building has recently been thwarted chiefly because of its historical significance.

Post-modern architecture

Post-modern architectural styles, when considered in the context of gentrification and heritage preservation, appear, first, to be part of a larger shift in approaches to making landscapes, and secondly, to be a late arrival. In this case architects seem to have followed a trend already set by commercial designers and histori-cal preservationists. They have adapted this trend to their own specific purposes by making it a self-conscious component of design. So Robert Venturi, writing in 1966 in *Complexity and con-tradiction in architecture* (p. 18), argued that architects should 'be guided not by habit but by a conscious sense of the past', because this is a primary source for the richness of meaning and ambiguity which makes built environments interesting and attractive. At about the same time the Italian architect Aldo Rossi began to formulate arguments for creating new urban forms based on the model of the pre-industrial European city with piazzas and a mix of formal and informal pedestrian spaces.

It was, however, not until the late 1970s that it became fashion-able and acceptable for architects to borrow consciously from past styles, to 'quote' visually Le Corbusier or some other master by making subtle references to his style, and to incorporate rounded arches and classical pediments or even elements of nineteenth-century workers' houses into their designs. The result was a frankly decorative and eclectic architecture. It came of age when Charles Jencks, an architectural journalist and critic, popularised a name for it — 'post-modernism' — a name which indicates clearly that its main feature is a deliberate rejection of unornamented modernism. Venturi had already found a neat slogan for it by turning Mies van der Rohe's 'less is more' principle for modernism on its head; 'less', he declared wittily (1972, p. 23), 'is a bore'.

Post-modernism is receptive to historical references of all sorts, but it is not much concerned with accurate reproduction. Bits and pieces from old styles can be mixed and matched, and architectural history is treated as a resource to be used in whatever way seems interesting and attractive. This is a cafeteria approach that permits visual quotations, metaphors, exclamations, and subtle references to famous buildings. Charles Jencks called his book *The language of post-modern architecture* for precisely this reason. It means that many post-modern buildings tend to be precious and self-serving, filled with allusions that only the architecturally super-educated can appreciate, and which to the rest of us look like an amusing hodge-podge of unrelated parts.

Recent though the development of post-modernism is the architectural details and features that are used are already quite standardised. Buildings are polychromatic, with trim painted in purples, blues, pinks and greens. Brickwork, widely used because of its associations with older buildings, often has distinctive quoins and dark and light stripes (William Morris called a nineteenth-century antecedent of this 'the streaky bacon style'). Half-round arches are such a common feature of post-modern buildings that they alone almost constitute an identifying mark for the style; presumably they are used because in steel frame and concrete construction such arches are pure decoration (Figure 11.6). Other fragments of past styles, such as round columns, pediments and pilasters are frequently employed for decoration; they are usually made of smooth concrete. And while modernist buildings had flat roofs, post-modernist ones, including skyscrapers, have gables and mansards.

These are the most common distinguishing features of what seems to be the most common style of post-modern architecture,

Figure 11.6: Post-Modern architecture. Post-modernism is a wide-ranging cafeteria style which either borrows freely from the past to create interesting ornamental effects, or adds commercial sorts of decoration to otherwise modernist buildings. The Markham Public Library in Ontario has both the rounded arches and the stylised classical columns that mark many historicist post-modern designs. The skyscraper with ornate railings and what seem to be multi-coloured garden sheds attached to its roof is the Vacation Inn in Christchurch, New Zealand. The face is the head of a building called 'The Turtle', the Center for Native American Art in Niagara Falls, New York, which has the form of a turtle; a geodesic dome constitutes its shell

one that Arthur Drexler has called 'historicist' because it is more or less historical with its references and allusions to gothic and classical orders. The Portland Building by Michael Graves and the AT&T skyscraper by Philip Johnson are possibly the two most famous examples of such historicism. There seem to be at least two other types of post-modernism, one contextual and the other a debased form of decorative modernism.

In contextual post-modernism new buildings replicate the main features of the surrounding structures without reproducing them exactly, maintaining rooflines, fenestration patterns, setbacks and so on. Robert Venturi has gone so far as to argue for an 'ugly and ordinary' architecture which borrows elements from commercial buildings and industrial state vernacular, including 'the technology of unadvanced brick, the old-fashioned double-hung windows', and television aerials, and then reworks these to create an expedient new building compatible with its suburban surroundings (Venturi, 1972, pp. 93, 102). On a much greater scale than this the architect François Spoerry designed a condominium cluster on the shore at Port Grimaud in the French Riviera, a cluster which looks for all the world like an old fishing village except that it is surrounded by yachts and cruisers rather than fishing boats.

Somewhat less common than the historicist and contextual post-modernisms is a version of it that attempts to transcend modernism by exaggerating, decorating and even ridiculing it. There is a skyscraper on John Street in Lower Manhattan which is a conventional skin-and-bones thing up above, but at street level has awnings, stage lights, steel picnic tables and chairs painted in primary colours, steel sculptures of bicycles in real bicycle racks, a complex electronic clock and an entrance through a tunnel of pink neon tubes. The Illinois State Center in Chicago is rectangular on two sides, with a stepped-back conical form on the other side enclosing a spectacular atrium and fronting on to a plaza demarcated by sculptural things that suggest broken columns; outside and inside the building is brightly coloured in patches of red and blue (Figure 11.7).

In the world of post-modernism almost anything is possible. There is a project by Stanley Tigerman for a suburban supermarket that is simply a greatly enlarged version of a suburban house, garage included, with a giant statue of a suburban housewife waving cheerily at the door. Buildings have been constructed which look as though great sections of the brick cladding have

Figure 11.7: More post-modernism. The State of Illinois Center in Chicago, completed 1985; the exterior has patterned panels of red and blue, and the T-shapes on either side of the Dubuffet sculpture are apparently stylised columns; inside there is a dramatic atrium, an essential element of all large 1980s buildings

crumbled or are peeling away. Such things suggest that post-modern architecture has deeply confused and multiple personalities. In fact it has a simple underlying schizophrenia out of which almost any combination of styles can be made. It is half rooted in the decorative qualities of the past, and half reaching out towards a fast changing, stylish, affluent society indulging itself in fashions for little more reason than that they look interesting. There can be little question that it is materialistic, superficial and arbitrary. At the same time it is clear that it is creating buildings and townscapes which are pleasanter to be in and of far greater visual interest than those infused with plain modernism.

Urban design

The planning equivalent of post-modernism is urban design, just as the planning equivalent of modernism was the institutionalised practice of planning by numbers. Urban design emerged in the late 1960s as a branch of planning which was concerned with giving visual design direction to urban growth and conservation (Barnett, 1982, p. 12). Whereas most planning is concerned with two-dimensional abstractions, such as subdivision layout and the segregation of land uses, urban design attends to the coherence of townscape, including heritage districts, the relationships between buildings both new and old, the forms of spaces, and small-scale improvements to streets — for instance, wide sidewalks, benches, attractive street furniture, provisions for outdoor cafés, and trees and landscaping. Much of the emphasis in such improvement is on visual quality, which is remarkable because this had been almost completely ignored in planning since the days of the City Beautiful movement at the beginning of the century. Of course, it also means that urban design can be criticised, as city beautiful was, for tending to ignore important social problems or papering them over with prettification, but this would not be a fair criticism both because urban design is no more than one element, albeit a particularly visible one, of a broad approach to planning, and because it is in practice often closely integrated with community revitalisation.

There are several means by which urban design operates. Development bonuses are used to encourage developers to include plazas or walkways or some attractive public space in a development in exchange for increased floor areas. Special zoning districts, in which locally specific design controls are applied, can be established — there is one for the Times Square theatre district in New York City; these may require, for example, pedestrian arcades or continuity of architectural style. After years of carefully segregating land uses there have recently been attempts to establish mixed-use zones which will maintain the variety of activities that are such an essential part of the grain and texture of the city scene. Site plan controls can be used to regulate the styles and forms of new commercial buildings. Standards have been established by some municipalities to ensure more attractive street furniture and signs than the utterly functional and unrelated benches and poles which generally prevail. And, rather than encouraging complete redevelopment of older districts, urban designers have generally argued, in concert with historical preservationists, for maintaining any streetscapes or buildings with visual and historical character, including old industrial buildings, by adapting them to new uses.

Much of the accomplishment of urban design controls has been preventive, for they serve chiefly to protect the public parts of cities from unscrupulous, individualistic developments; in this respect their consequences are not easily identifiable even though they may be considerable. Their specific landscape effects are patchy, which is to be expected given their recent implementation, and are often closely integrated with heritage projects and commercial gentrification. Thus Ghirardelli Square in San Francisco, where a former chocolate factory was in 1964 converted into a shopping centre, is often considered a prototype urban design project. Clearer examples of landscapes created by urban design are the pedestrian arcade on the east side of Lincoln Square in Manhattan, the slender apartment buildings constructed on the crowns of the hills of San Francisco in order to emphasise the relief and to preserve views, pedestrianisation projects like that around Leicester Square in London, the carefully articulated housing clusters created under the design guidelines for residential areas set by the Essex County Council, and the construction of new suburban plazas in Toronto at the street line rather than set back behind barren asphalt parking lots (Figures 11.8 and 11.9). As long as urban design controls continue to act as the planning link between gentrification, heritage districts, architectural design control and community revitalisation, these fragmentary manifestations can be expected to grow and connect,

Figure 11.8: Urban design guidelines for San Francisco, and for Essex in England, illustrating how slender buildings reinforce the shapes of hills and bulky ones do not, and the need to group buildings in new urban areas in order to avoid 'the visual failures of recent housing developments'

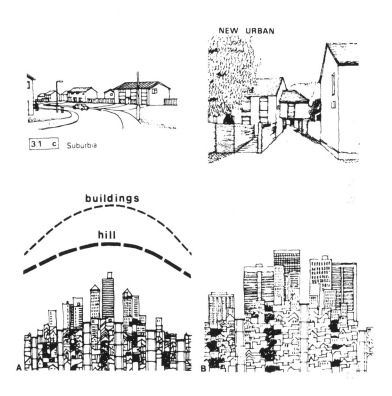

Sources: San Francisco, 1971, p. 80; Essex County Council, 1973, p. 62

and thus eventually to have a widespread effect on urban landscapes.

Community planning

Urban design, heritage preservation, post-modernism and gentrification have been paralleled by a growing orientation to community-based planning. Whether this is just coincidental or stems from some single underlying cause is not clear. The civil rights, anti-war and student protests of the late 1960s, as well as

Figure 11.9: Landscapes of urban design. In seeking to improve the visual quality of cities urban design has often turned to old street features and adapted them to the modern environment, and to that extent it is the planning equivalent of post-modern architecture. The photographs show a sight-line on to a cathedral in Toronto opened up through the application of urban design principles, and a fragment of the revitalised town centre in Burlington, Ontario, with many of the detailed motifs of urban design — old-style street lamps, benches, planters to shield the parked cars, and the pedestrian crosswalk identified by different surface materials

protests against urban renewal and expressway construction, raised political awareness, and there is no doubt that middle-class gentrifiers are anxious to and capable of articulating their concerns about urban environments. For some combination of these reasons a significant change occurred in urban planning during the 1970s. The relatively authoritarian, planning-by-numbers approaches, which had been in use since the late 1940s, were joined by a community planning process aimed at developing solutions through consultation and neighbourhood workshops. Mark Francis has listed the differences between the two methods. Community planning is small-scale, local, includes the users as clients, is low-cost and democratic; modernist planning is large scale, uses national or international models, corporate oriented, high-cost, and its solutions are imposed by experts who assume that they know what is best for residents. This is probably overdrawing the distinction, but the fact remains that some degree of community involvement in planning is now commonplace, and indeed the opportunity for it is now legally guaranteed in many municipalities.

In its most completely developed forms community planning involves a large degree of citizen participation and self-help. For example, Randy Hester co-ordinated a plan for Manteo, an economically depressed island community off the North Carolina coast, in which he asked the residents to identify the places which were special to them (these included the post office and many 'homey and homely' places as well as the marshes, shoreline and public monuments), and then drew up a development plan to protect these as much as possible while encouraging some economic improvement. Local volunteers and craftsmen constructed a boardwalk which was intended to promote tourism, and the overall plan has been implemented using modified zoning controls to protect the special places. Lucien Kroll adopted a not dissimilar approach in designing student housing at the University of Louvain in Belgium, and Christopher Alexander and his colleagues co-ordinated a development plan for the University of Oregon which is based on active and continuing involvement by all members of the university community including professors, students, janitors, secretaries and administrators.

This degree of community consultation and involvement in design is still uncommon (and there are, in practice, many difficulties with non-participation and vested interests). However, it is now not unusual for planning departments to have neighbour-

233

hood planners working from storefront offices, or to provide opportunities for community consultation concerning development proposals. In California and Britain, and no doubt elsewhere, there are government programmes to promote and assist partially self-built housing developments. In Massachusetts a number of financial corporations, such as the Massachusetts Commercial Development Finance Corporation and the Greater Boston Development Corporation, provide funds and management expertise for promoting neighbourhood improvement, including the revitalisation of commercial areas and small industries and providing housing for low-income groups, the handicapped and the elderly. Local labour is used wherever possible. Such community-based, bottom-up planning and development is now sufficiently commonplace to be unremarkable. What is remarkable is that there was almost none of this 20 years ago.

The landscapes of community planning and participation are rarely exceptional. They might be upgraded old row houses, renovated stores, new stacked townhouses or ten-storey apartment buildings. They are not pretentious or flashy. During construction there may be signs which tell strangers that this is a self-help project, and an astute observer may sometimes be able to guess from the scale and character of renovated or new developments that they are a product of community involvement. The edges are blurred, the land-uses mixed, the buildings almost self-consciously unassuming. Left mostly to their own devices people tend not to segregate activities and to draw sharp boundaries. They prefer buildings that fit into their context and patterns of activities that are as complex and fuzzy as everyday life.

There are, however, a few exceptional places which have been created by community involvement. One of the most important of these, almost a prototype planning development, is Village Homes at Davis in California (Figure 11.10). This is not strictly a community planning project in that it was conceived, promoted and built by a local developer, Michael Corbett, but it is a development whose success depends largely on the co-operation and participation of the residents. It is significant because it offers an alternative to the conventional models of subdivision development that are now so widely used. Village Homes is an energy efficient, environmentally sensitive co-operative. Its streets are laid out as a series of east-west culs-de-sac for the houses to take maximum advantage of passive solar heating, and many build-

Figure 11.10: Village Homes in Davis, California. Developed in the 1970s, Village Homes is unusual partly for its community planning (though much of this came from the enthusiasm of the developer, Michael Corbett, for community involvement), and partly for its innovative physical form stressing energy efficiency. The plan shows part of a cul-de-sac oriented east-west for maximum passive solar gain, internal pathways and bikeways, and houses clustered around a communal garden. The photograph of such a cul-de-sac shows that these are substandard streets with no sidewalks, visitor parking is accommodated in off-street bays, and many of the houses have cultivated yards rather than lawns

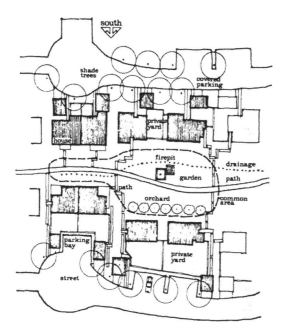

ings also have solar panels. There is a natural drainage system with channels that allow surface run-off to seep away instead of flowing into storm sewers. The streets have substandard widths, there is no on-street parking and parking bays for visitors are provided, so that much less land than normal is devoted to vehicles. A system of internal pathways/bikeways links the whole development. The houses are clustered so that each group shares a communal vegetable garden. There are few lawns or open spaces except for a communal recreational area. There is also a communal vineyard and orchard, part of the latter on what would normally be the boulevard by the side of the road.

Village Homes is clearly in the garden city tradition, that is to say it is imaginative and has received much attention, and maintains the emphasis on community and co-operation which has characterised many of the best planning ideas of the twentieth century. Like the garden cities it is unlikely to be widely copied except in fragments because co-operation is not easily reproduced through physical forms. Nevertheless Village Homes demonstrates the possibility of suburban development forms which are very different from those which are conventionally used around the developed world, and it breaks away from the over-engineered standards for roads and drainage which have contributed so much to the uniformity and placelessness of modern suburban landscapes.

Late twentieth-century eclecticism

From their combined effects on landscapes it is not easy to draw significant distinctions between gentrification, heritage preservation, post-modern architecture, urban design and participatory community planning. Projects like the revitalisation of Covent Garden in London, for example, or the fake Mediterranean fishing village at Port Grimaud in France, do not fit neatly into any one of these categories, nor do they fit easily with any older conventions. Since about 1970 architecture, planning and business corporations have all moved in a similar direction by encouraging greater variety, old as well as new styles, and responsiveness to public opinions in landscape making. Of course there are deeply entrenched habits of thought and ponderous institutional practices to be overcome before these new methods can be said to prevail. Nevertheless they have already caused the collapse of the single-minded modernist vision of a future urban landscape filled with great skyscrapers, megastructures and machines, a landscape which would have been an austere celebration of scientific technologies and rationalism.

In the era of post-modernism almost any type of urban landscape has become both possible and popularly acceptable — old and quaint, new and quaint, modernist, decorated modernist, flashy, hi-tech, sleek, vernacular, or frankly fake. Skyscrapers and modernist projects will continue to be built, but they will be joined by victorian village shopping malls, self-built housing and exotic post-modern buildings in a multitude of different styles. It is already clear that the new urban landscapes of the last years of the twentieth century will be distinguished above all by their eclecticism. This is not without its problems. When buildings and places can take on any appearance one chooses, and there is rarely any substantial reason for choosing one appearance over another, it is easy to fall into a paralysis of indecision that can be overcome only by making arbitrary design choices. The consequence in the near future could easily be urban landscapes which are a chiaroscuro of increasingly flashy, unrelated and pointless patches, a post-modern, late-modern monotony-in-variety.

12

Modernist Cityscapes and
Post-Modernist Townscapes

'The architect', W.H. Whyte has written (1982, p. 26), specifically about the Seagram Building in New York but he could have been refering to anywhere new, 'sees the whole building — the clean verticals, the horizontals, the way Mies turned his corners, and so on. The person sitting in the plaza may be quite unaware of such matters. He is more apt to be looking in the other direction.' And what he sees in the other direction is not likely to be architectural masterpieces standing in splendid isolation, but a layman's view of a street with buildings, signs, trees, parking spaces and people, all together. This is 'townscape'.

Townscape, the view of the street, is what most of us experience of cities most of the time. Streets may seem obvious because they are so commonplace, but in fact they have a complex content of structures, materials, things such as art objects and benches, cars and pedestrians, as well as the spaces of the roadway and the sidewalk. Townscape, wrote Gordon Cullen, the English architectural journalist who developed the idea in the early 1950s, is about the 'art of relationship' between all of these (1971, p. 7). It recognises above all that our experiences of towns and cities are not so much studied contemplations of a single view as encounters in passing, unfolding sequences of street scenes. Cullen called this dynamic experience of townscapes 'serial vision'.

These sorts of ideas must have been in the air in the 1950s, because at about the time Cullen was examining townscapes in Britain, Kevin Lynch in America was writing about what he called the 'sensuous form' or perceptual coherence of urban landscapes — their spaces, the diversity of sensations they provide, their vitality and sense of place, and their sequences of views.

238

Townscape and sensuous form are perceptive and powerful ideas. On one level they challenge the specialised and fragmented perceptions of architects, planners, developers and engineers by drawing attention to the totality of urban landscapes. On another level they seem to grasp the essential qualities of places which many people seek out as though instinctively when they have the freedom to do so, for instance as tourists. Given half a chance people do seem to appreciate quaint old townscapes packed with sensuous form, like those of Venice which Lynch used to describe the merits of an interesting sequence of spaces, activities and textures. And here we can begin to grasp the problem of modern urban townscapes, because an interesting and coherent sequence of spaces, activities and textures for strolling pedestrians is an exact description of what most of them do not have.

Design phases in twentieth-century townscape

The development of townscapes in the twentieth century falls into three phases. Now townscapes, like all landscapes, are ponderous things, slow to change and even slower to react to bright new ideas about how the world should look. Furthermore, not everywhere is inclined to react to changes at the same rate. So these three phases are not equally clear in all townscapes — city centres, presumably because of their high property values and intensity of use, seem to reflect changes first, whereas some small town main streets may still be little changed from the early decades of the century. Nevertheless three broad phases can generally be recognised.

The first, transitional, phase lasted up to about 1940 and was marked by incremental changes to older urban forms as new technologies and concepts were introduced. Most of the streets made in this period, except for those in Manhattan, had forms and styles which would not have been incomprehensible, say, to Samuel Johnson. Accommodations were, of course, made for automobiles, but these were changes of detail and even when the streets were entirely new they were lined with three- or four-storey buildings, and had the sort of scale and decoration that made sense to pedestrians. In short, street form remained much as it had been for centuries (Figure 12.1).

Of course, behind the façade of this quite traditional street form all sorts of changes were occurring, changes which ranged

Figure 12.1: Twentieth-century streetscapes: Golders Green High Street, London, about 1930, built in the early decades of the century, not unlike streets made anytime in the previous 500 years; a modernist street in suburban Rotterdam, apartment buildings set back from the street line and the roadway devoted to vehicles rather than pedestrians; a post-modernist street in a shopping plaza in Aurora, Ontario — pedestrianised, intricate spaces lined with fake medieval buildings

Source for Golders Green: Clunn, 1932, p. 388

240

from the development of steel frames and socialist ideals, to the growth of corporations and the invention of methods for town planning. These only manifest themselves in something more than a fragmented way in the decades after World War II. This was the second phase of townscape development. It involved the dramatic changes that accompanied modernist architecture, corporate development and institutionalised planning. It reached a zenith in the 1960s and 1970s and continues to the present, though with diminished intensity. In Le Corbusier's unequivocal expressions, modernism sought to create entirely new urban forms, and to kill the street, to make the street into a traffic machine. To a considerable extent it succeeded in these aims, with one result being that whatever Cullen and Lynch wrote about townscape has almost no relevance to understanding modernist cityscapes. These have rational not sensuous form, street spaces are deep and straight, individual buildings are designed without concern for context, their surfaces are barely decorated, there is little texture, visible activity has for the most part retreated inside the buildings, and serial vision is primarily that of the driver — with all that entails about reduction in details and shifts in scale.

Fortunately this is not the end of the story. Since about 1970 streets, or to be more accurate some bits of streets, have been rediscovered and reclaimed for pedestrians, albeit in a rather contrived way. This third phase is one of the development of post-modern townscapes. Whether this has happened because theoretical notions about townscape filtered through into practice, or because of a reaction against the manifest deficiencies of modernist machine streets, or simply because of a shift in fashion, is not entirely clear. What is remarkable is that it came about so quickly. Modernism was brewing for 50 years before it began to rework urban landscapes on a large scale; post-modernism seems to have had a gestation of less than a decade before its effects on cities became apparent. And it is now clear that heritage planning, urban design and post-modern architecture, taken together, have contributed to a revival of interest in the character and quality of traditional streetscapes and that a distinctive, if fragmentary, post-modern townscape is being widely created.

In a model case the landscapes of all these phases would be found juxtaposed. There would, perhaps, be a modernist city core of skyscrapers and canyon streets, surrounded by post-modern districts of gentrified housing and warehouse/boutiques, then some 1920s garden suburbs with traditional retailing streets surrounded in turn by a band of corporate suburbs laid out in neighbourhood units, split by arterial roads and shopping malls, and then a ring of sleek late-modern office buildings and electronics plants lining expressways or grouped into urban villages. In large metropolitan areas it may indeed be possible to identify a landscape pattern not unlike this. In smaller cities and towns it is, however, entirely possible that one or more of the phases is missing. Most new towns, for example, are wholly in the institutionally planned, ordinary modernist style of the 1950s and 1960s; they have no main streets, only roads and plazas and pedestrian precincts, and they have no old buildings and districts to preserve or gentrify. They have, in fact, a one-dimensional landscape. Conversely, in old small towns it is entirely possible that the urban renewal and growth of the high modernist phase has passed them by, and they have skipped directly from traditional street forms to the heritage projects and streetscape revitalisation of post-modernism. This is especially obvious in places, such as Carmel in California or many of the hilltop towns of southern France, which have recently been rediscovered for their quaintness and turned into artisan or retirement communities or tourist attractions.

Qualities of modernist cityscapes

Writing generally about modern construction, Marston Fitch claims (1961, p. 229) that engineers have 'raised the order of magnitude to literally geological proportions'. This is an appropriate image for the modernist city centre where an entire desert physiography has been created — with canyons, sheer cliff faces, often empty spaces, hard surfaces hot in summer and cold in winter, windstorms with swirling dust and debris, and infrequent oases of vegetation. From the distance of a passing freeway most of this is not immediately apparent; all one sees is a great mass of towers, solid and gleaming during the day and at night a spectacular display of disembodied lights.

In the suburbs an entirely different profile presents itself, one of open spaces, with wide streets and buildings set well back from the street line; perhaps an occasional cluster of taller buildings serves to emphasise the horizontality of it all. Residential roads are trimly landscaped. Along arterial routes the landscape is windswept and barren, and even though there are sidewalks few pedestrians venture into this world of exhaust fumes, crushed hamburger boxes and brown ribbons of broken cassette tapes.

There is nothing in either of these typical modernist settings that evokes anything to do with the small scale and detailed textures of towns, so I will call them 'cityscapes'. They are set apart from both pre-modern and post-modern urban landscapes by their bigness, the shapes of their spaces, their rational order, hardness and opacity, and the character of the serial vision which they provide.

Megastructural bigness

Modernist built-forms are, to put it very simply, bigger, taller, wider, than almost all their predecessors. They are, in fact, megastructures. Actually Reyner Banham (1976, pp. 1-3) defines megastructures as being modular, extensible, prototype city structures conceived in the 1960s and 1970s; this seems unduly restrictive since the word so clearly refers to any very large building, and since there are so many very large modern buildings in need of a word to describe them. So I consider a megastructure to be any very big structure, whether vertical or horizontal; it could be a single building or a group of buildings constituting a single development. Megastructures are the visible expressions of administrative megamachines, so they often occupy entire city blocks and dominate their surroundings either by height or by sheer bulk. The Pentagon is a megastructure by this definition, and so are the World Trade Center, the Rialto Project in Melbourne, the Barbican in London, the Houston Astrodomain, the Renaissance Center in Detroit, La Defense in Paris, and Los Angeles airport.

When inserted into existing townscapes megastructures are distinguished by the fact that they have few street level entrances and little in the way of details. The nineteenth-century street was based on store-front units of 20 or 30 feet; the megastructural street has units of 200 or 300 feet. The result of this is a special and almost universal effect: the street is rendered lifeless and pedestrians avoid it if at all possible. This happens wherever there are skyscraper offices, convention centres, new institutional structures, or even modernist housing complexes.

Straightspace and prairie-space

In the canyons of the new city centre straightspaces predominate; streets are much deeper than they are wide, with wells and clefts between buildings. These are bounded by the flat surfaces, straight lines and sharp edges of curtain walls, with perspective lines vanishing towards the horizon. In contrast, in new suburbs prairie-spaces, characterised by a sort of horizontal vacuity, prevail. Wide roads are lined by parking lots and low, no-frills modernist plazas, or by neat houses (Figure 12.2).

The spaces of the modernist cityscape tend towards these two extremes, but many are left stranded in an intermediate netherworld of 'spaciness'. By this I mean spaces which are ill-defined, neither deep nor wide, neither clearly enclosed like those of the medieval town nor carefully proportioned like those of the renaissance city. They are awkwardly lacking in attractive or sensual properties. Spaciness describes spaces and places which look as though they are what was left over after the buildings and poles were stuck up (Figure 12.3). It is particularly common around high-rise apartment buildings. There may be a lot of room but it seems to serve little purpose or to have no attractive form; it is often landscaped in a half-hearted way or turned into parking. These spacey areas are rarely pleasant for pedestrians to be in, to sit or stroll in; they are to be hurried across on the way to the car or another building.

Rational order and inflexibility

Modernist designs are totally designed packages which allow almost no scope for additions or modifications. Unlike traditional architectural styles there are few doorways or windows that can be altered, no trim to paint a different colour and no decorations that can be easily changed. Styles are designed in detail and intended to last for the lifetime of the building. The aim of this seems to be part of a larger process which requires that everything be put, and then maintained in its proper place so that order and neatness prevail. Hence lamp-posts, trees, benches, apartment blocks in urban renewal projects, are arranged in rows or in self-conscious clusters. Uses and activities are carefully segregated at all scales, from industrial and residential districts down to the eating areas in shopping malls.

Figure 12.2: Spatial extremes in modernist townscapes: the straightspace canyons of city centres, and the prairie-spaces of the suburbs

Figure 12.3: Examples of spaciness. This is not easily conveyed in photographs, but the empty and awkwardly bounded concrete space of Paternoster Square in the City of London, and this fragment of suburban Houston, perhaps give some sense of it

It is easy to imagine and to understand the rational mind at work behind the rational order of modernist landscapes, laying out plans with a set-square on graph paper, considering space allocations, accessibility, turning circles for vehicles, emergency fire routes, illumination and maintenance standards; attempting, in fact, to anticipate all eventualities so that nothing untoward shall ever happen, safety is ensured, efficiency is guaranteed, and the form of the future city is predetermined. These are sound motives, hard to dispute. Their unfortunate consequence is the elimination of uncertainty and excitement from the urban scene, and the simultaneous imposition of inflexible forms and standards. As if to make this explicit, standardised official signs informing us what we are not permitted to do sprout from every post. No Parking, Do Not Walk, No Entry, No Access, No Trespassing, No Stopping, Post No Bills, No Littering, No Loitering.

Hardness and opacity

In the modernist cityscape materials are hard. Perhaps the resulting straight edges have a modernist look, or perhaps these materials are easier to maintain and less susceptible to vandalism. Whatever the case, hardness reduces textures, variety, and the possibilities for public involvement in design. It can be found at almost every scale in almost all modern landscapes. Robert Sommer, an environmental psychologist with a good eye for these sorts of things, writes (1974, p. 2) that 'the hardening of the landscape is evident in the ever growing freeway system, the residential and second-home subdivisions pushing aside orchards and forests, the straightening and cementing of river beds, the walled and guarded cities of suburbia, and the TV cameras in banks and apartment buildings.' Its particular manifestations are opaque forms, a lack of details and prison-like concern for control (Figure 12.4).

The blank concrete walls and reflecting glass windows of megastructures reveal little of what goes on within. We see only water stains on the concrete and reflections of ourselves or of other buildings. Many of the doors at sidewalk level are blank fire doors that only open outwards. Visual and physical access is everywhere restricted. Opacity prevails. Inside they could be making Cabbage Patch dolls or delivery systems for nuclear weapons. We can't tell.

Figure 12.4: Hardness, lack of detail and rectangularity in modernist townscapes: a corner of the Dallas Convention Center (designed, according to a plaque set into the building, by a group called Omniplan), and Miesian benches designed to accord with modernist aesthetic theory rather than comfort in the plaza of the Toronto Dominion Centre

A hallmark of modernism is the lack of ornament and decoration. In modernist cityscapes this means that there is no hand-made detail, no evidence of craftsmanship to draw and to hold our attention, except perhaps for the occasional graffito on a blank wall. Colours range through a limited palette of pale greens and browns to metallic and concrete greys. Textures are smooth like glass, or rough like concrete aggregate, in neither case especially pleasant to the touch. Put bluntly — at street level unornamented modernism is boring.

The aesthetic inspiration behind hard cityscapes may once have had something to do with creating a style appropriate for machines and mass-production. This justification now seems historically remote and irrelevant. Sommer argues that the real model for modernist urban design seems to be the prison — everything spartan, secure, easily maintained with barren open spaces, metal fences and blank façades, authority evidently in control. The assumption behind this seems to be that if something can be damaged it will be, so it has to made out of concrete or metal to be as damage resistant as possible. A second assumption is that people tend to behave irresponsibly and unsafely, so fences and barriers and signs must be erected to reduce this possibility. The widespread results of these assumptions can be easily encountered in almost any modernist cityscape, especially by exploring those parts where ordinary pedestrians are not expected to look, around the back, in the service areas, and so on. Delivery bays in plazas, public housing developments, schools, airports, expressways, are invariably demarcated by chain link fences and prohibitive signs, and made of virtually vandal-proof materials, just one step removed from a detention centre.

Discontinuous serial vision

Serial vision in modernist cityscapes is a matter of views and perspectives separated by sharp disjunctions. Consider driving into a city. From the swooping curves of an expressway the skyline of skyscrapers is a spectacular declaration of the status accorded to corporate dedication to conspicuous administration. Down the off-ramp, into the street canyons and the driver has an altogether different perspective, one that is nervous, stop and go, crowded with other vehicles and pedestrians, the magnificent tops of the buildings out of view, their bases disconnected and hard to see since attention must be directed ahead, signs to watch for and signals to obey. It is a relief to turn into the concrete caves of underground or multi-storey garages, there to be transformed into a pedestrian, even if it is the case that of all experiences of modern urban landscapes that of the pedestrian in a parking garage is undoubtedly the dreariest. They are functional to a fault, dusty, dim, with low ceilings; one can only scurry to the nearest stairwell and hope nothing malevolent awaits.

The pedestrian's serial vision in the modernist cityscape is scarcely less disjointed. It is comprised of vistas vanishing like the lines in a perspective drawing, interspersed with right-angle turns which abruptly reveal new views not unlike those which just vanished from sight. There are discontinuities, too, in moving from grey exterior spaces to the bright interior spaces of atriums and malls filled with colourful signs and plants and people. There are few gradations or anticipations. It is an either/or world. Either straight ahead, which view is totally revealed, or perpendicularly off to the side; either outside in the machine street or inside in the pedestrian atrium. The only exceptions to this discontinuity are the disembodied tops of skyscrapers which can sometimes be glimpsed in the distance and which can serve as landmarks. Yet as one draws closer these become increasingly difficult to see without an unnatural craning of the neck. And when we finally reach their base the scale and appearance they present may seem so unlike the top, which is now impossible to see, that the two bear little relationship to one another (Figure 12.5).

The sleek styles of late-modernism have done nothing to mollify modernist landscapes. Indeed the mirror glass, grand structural displays and concrete boxes have carried opacity and hardness to new levels of space-age sophistication. The crystal clusters of the Renaissance Center in Detroit and the Westin Hotel in Los Angeles gleam spectacularly from a distance, but people are not meant to walk to them so the main entrances are grimy automobile loops and underground parking ramps; pedestrians are greeted by concrete walls and flights of steps which seem to lead nowhere. The sleek late-modern megastructures that corporations have built as their suburban headquarters are no less inflexible and hard; they stand alone, fortresses isolated by their landscaping and parking lots, protected by their video security

Figure 12.5: Discontinuity and connection of serial vision in modernist townscapes: the impressive mass of a city skyline of skyscraper offices is visually scarcely connected with the street canyons that offer distant glimpses of individual skyscraper tops, which are, in turn, scarcely related to the views of skyscrapers at their base. The skyline of mid-town Manhattan from Central Park, with the post-modern top of the AT&T skyscraper at centre right; the same building seen at the apparent end of a street canyon; and its base in a different style

cameras. Nobody walks to them or around them except the grounds maintenance staff.

Such buildings are consistent with a society dominated in many of its aspects by automobiles. They are either to be glimpsed as shimmering towers and elegant shapes from the distance of an expressway, or they are to be experienced on the inside where their plant-festooned atriums soar upwards. One is either outside in a car or inside on foot; the transition between the two is apparently meant to be achieved by some sort of quantum leap, so the doorways, elevators and passages linking parking to atrium are drab and uninspiring. For those who choose occasionally to walk outside this means that the late-modern city street is no less dispiriting than the modernist one. Both are equally placeless and international, equally futurist in their suggestions, equally renitent and unyielding towards pedestrians. Though they may have to work here these are not landscapes in which many people wish to linger. They are to be hurried through on the way to districts which still retain some suggestions of local character and human scale, or where post-modernism has begun to unwork the unrelenting hardness and sheen of the modernist cityscape.

Qualities of post-modern townscapes

In the late 1960s there were few new exceptions to the prairie spaces and hard canyons of the modernism, and the tacit assumption then seemed to be that any remaining older landscapes would eventually be remodelled into these sorts of forms. This situation has changed. Modernist cityscapes are still being built, but so too are post-modern townscapes with the appearance of the hand-crafted qualities, detailed textures and intricate spaces that Cullen and Lynch so admired. These townscapes are the product of the combined effects of heritage preservation, urban design, post-modern architecture, gentrification both commercial and residential, and community planning. Since these are such recent events in the development of the modern urban landscape their overall impact on the look of cities is as yet limited. Nevertheless, there is no question that there has been a major rebound from the dreary grey spaciness of modernism. The 1980s have witnessed a celebration of differences, of polyculturalism, of variety, of style and stylishness, and post-modern townscapes are a clear expression of this celebration.

A process of deliberate diversification now seems to be at work in many cities, a process which saves, celebrates and yet integrates into the existing social order almost everything to do with visible differences. Ethnic identities are emphasised in festivals, in restaurants, in such things as Greek/English or Chinese/English street signs, pagoda-style phone booths and multi-lingual signs on the outlets of multi-national corporations. Local history has been rediscovered, monuments restored, pioneer villages constructed, historical districts designated, almost defunct festivals and traditions like Morris dancing have been revived; anniversaries, centennials, bicentennials, and millennials of the foundings, discoveries and first settlements of cities and nations are celebrated everywhere. The decoration of interesting public spaces for their own sake has become commonplace, with hanging baskets and planters for flowers, and attractively designed rather than merely utilitarian street furniture and signs. Urban design and pedestrianisation projects have revitalised old streets.

The post-modern townscape is the visible, built expression of this web of social and cultural changes which are changing the character of cities. Its distinctive features include quaintness, textured façades, stylishness, attempts to establish connections with local history or geography, and a clear split between pedestrians and automobiles.

Quaintspace

The overriding characteristic of post-modern townscapes, especially in comparison with their modernist predecessors, is quaintness. It is in fact probably legitimate to talk of them as products of a process of 'quaintification'; they have quaintly named stores, such as The Hobbit or The Olde Village Shoppe, they have textured surfaces made of materials like 'heritage brick', they have a quaint medieval sort of scale, and they have quaint spaces.

Quaintspace involves intricate sequences of enclosures, winding passageways, little courtyards, canopies over sidewalks, easy transitions and continuity of appearance between exterior and interior spaces (Figure 12.6). There are few right angles, views are deflected with frequent suggestions of interesting things up ahead. In the courtyards and squares there are often provisions — stages, lighting, and so on — for street theatre, little concerts or displays. No doubt all the performances have to be officially sanctioned, and the performers have to obtain suitable licences (this is, after all, the late twentieth century), and to that extent

Figure 12.6: The quaintspaces of post-modern townscapes: a revitalised back alley in Santa Barbara with intricate pedestrian spaces, comfortable benches, textured surfaces, canopies and flags, planters with flowers and trees, half-round windows, and olde worlde lamp standards

there is contrivance in their apparent spontaneity. Nevertheless, quaintspaces seem to provide settings which encourage such activities, and in them there is ample opportunity for what Cullen calls 'the possession of territory', people engaged in specific activities in particular places.

Textured façades

Presumably because they are designed for pedestrians, post-modern townscapes are rich in the details of signs, materials and decorations. Surrounding buildings are likely to be of pre-megastructural dimensions, less than four storeys, and these old street forms are replicated. Pre-modern materials, including brick, barnboards and wrought iron, and pre-modern forms and styles are preferred for almost everything. Street furniture consists of ornate lamp standards with a late-victorian look, park benches, and sensitive landscaping with shrubs and flowers. Interlocking bricks and granite cobblestones are used for sidewalk surfaces, and to indicate pedestrian crossings on traffic streets.

This is, it should be noted, all on the surface and on the public side of the landscape. Careful inspection may reveal that the old-looking brick buildings are faced with a cladding less than an inch thick. Uniform air-conditioning units and meters for electricity remind us that for all the aged appearance this is a new environment. And the backs of the buildings, where the deliveries are made and the garbage is picked up, often have all the hard forms of modernism.

Stylishness

Post-modernism, as the most recent expression of urban environmental taste, is undeniably fashionable. It has a commercial chicness to it. The freshly cleaned and sandblasted surfaces, carefully hand-made signs, designer products, little art galleries, are all a testament to this. Even multinational chains participate, and McDonald's and Kentucky Fried Chicken try to merge into the local ambiance. Perhaps these new-old places will weather and mature, but for the moment they have all the self-consciousness of a media personality in public, always expecting to be the centre of attention without being too obvious about it.

Closely related to this stylishness is the fact that many post-modern townscapes are filled with more or less subtle indications of affluence (Figure 12.7). Here consumption is not gauche and flashy, perhaps because the heritage styles and quaintness tend to restrict the possibilities for ostentation, but more probably because post-modernism aspires to be the setting for a discriminating class of consumers. The signs of affluence and pretension

Figure 12.7: Indications of style and affluence: Rodeo Drive in Beverley Hills has a sheen of glamorous perfection to which many retailing districts of the 1980s aspire. Han Walm's is on Yorkville Avenue in Toronto

are nevertheless detectable. Polished marble trim, chrome and brass nameplates and signs, designer fashions in the stores, expensive antique furniture, no prices in window displays, the celebration of celebrity (most obvious in restaurants called Bogart's, Monroe's, Gable's and so on), signs listing cities where other branches of the store are located (Sassoon's of London, New York, Paris, Tokyo, Rome). So for all their oldish appearance and adoption of local historical motifs, post-modern townscapes are scarcely less international than those of modernism. Certainly they have sprung up almost everywhere at once and seem to cater to a wealthy jet-set clientele, though this is a clientele which apparently appreciates local environment character, even when it has to be contrived.

Reconnection with the local setting

Many post-modern townscapes re-establish contact with rivers, lakefronts and old industrial and residential districts that have long been ignored as amenities. In the nineteenth century rivers and lakes were used mainly for raw waste disposal, and lined with factories and railways. Now these are being reclaimed, canals are landscaped, riverside and lakefront walks are created, old industrial buildings are renovated into malls, expensive apartment complexes and convention facilities are sometimes squeezed in beside former warehouses, tourists appear, a post-modern district thrives. This is not historical restoration so much as historical reversal, or, to put it more charitably, an elaboration of the old that undoes history in order to create an entirely new sort of environment which has an historical and natural feel to it. Notable examples include Georgetown Mall in Washington, DC and Covent Garden in London, though almost every city has an equivalent district (Figure 12.8).

Pedestrian/automobile split

Modernist cityscapes put the automobile before people, relegating pedestrians to narrow sidewalks, tunnels and malls. Post-modern townscapes go some way to redressing this imbalance. They are primarily pedestrian settings — cars are either kept on the outside or their access is severely restricted. There is a degree of schizophrenia in this because automobiles are so essential to modern urban life and to the success of post-modern townscapes. At its most obvious, for instance in shopping centres like The Barnyard at Carmel in California, this schizophrenia results in post-modern townscape fragments which are surrounded by

Figure 12.8: Reconnections with the past: Georgetown in Washington, DC. An aspect of the post-modernising of townscapes has been the discovery of the visual possibilities of previously discarded landscapes; in this case the Chesapeake and Ohio canal has been reclaimed as an urban amenity and the adjacent warehouses have been converted into shopping malls and expensive condominium apartments

parking — quaint islands in asphalt oceans. The split is less apparent where post-modernisation has involved the revitalisation of an old area of a city. In these cases the pedestrian streets and squares, wider sidewalks, and people in the street seem to be the most obvious facts, but street furniture and spaces are usually so arranged that emergency and perhaps delivery vehicles can have access whenever necessary.

A modernist or post-modernist future landscape?

Perhaps post-modernism in all its varied forms is the harbinger of a new type of urban landscape, one produced through locally-based community planning and careful urban design, and one in which people take increasing responsibility for the appearance and the form of the places where they live and work. I would like to believe that this is going to happen. From a more cautious, if cynical, perspective, however, it seems that in spite of their recent diversity landscapes are increasingly pre-mixed, ready-made and imagineered, and it may be that post-modernism is little more than a disguise for ever more subtle and powerful types of rationalistic organisation by corporations and governments alike. For all its pre-modern suggestions of quaintness, its apparently old materials and its revival of locality, the post-modern street is still usually a product of large-scale economics and intense design efforts. Its appearance is the consequence of arbitrary choice and fashion rather than tradition, and is rarely more than superficial. To that extent it is a lie, though it is a pretty lie and one that seems thus far to have been widely successful in attracting pedestrians, revitalising decaying districts and contributing to the creation of differentiated urban landscapes. We can counsel perfection and condemn such deceits, but it is wise to be aware that not all post-modern townscapes are deceptions, and also that the only current alternatives appear to be the aesthetically honest yet street-deadening forms of modernism or the sleek opacities of late-modernism.

Regardless of whether we prefer community design, or the pretty lines of post-modernism, or the hard truths of modernism, in the near future urban landscapes will continue to be dominated by the constructions of the recent past and by deeply entrenched practices and habits of thought about the merits of corporate developments and municipal planning. The enormous changes made to the look of cities over the last 50 years will not soon be undone, they represent too great an investment of time and money for that to happen. Unless some fundamental change occurs in the social and economic order, and there are few indications that this is about to happen, the primary fact about future urban landscapes will be that, like most of those made since about 1940, they will have been designed, planned and created for rather than by the majority of citizens, and for reasons of efficiency and profit rather than community or social justice. Their forms, whatever they may be, will surely reflect this.

A unity of disunity — concluding comments

In his book *Looking at cities* Allan Jacobs ably demonstrates that walking is the only way to notice the wealth of detailed clues about social activities that cities offer us. Only by walking around a city can we see such things as the array of doorbells that reveal the number of apartments in a building, notice the superficiality of gentrification or the lack of decoration on modernist mega-structures, and experience the full climatic acceleration effects of skyscrapers. On the other hand, probably the only way to come to terms with the overall landscape character of a city in the late twentieth century is to drive around it. Driving makes it possible to grasp the huge spatial scale, the large patterns and the segregation of activities that are such major aspects of modern urban landscapes.

To drive around a city in the 1980s is to encounter a limited range of different types of townscapes, indefinitely repeated. These are, in fact, so different that they seem to bear little or no relationship to one another. There are drab modernist renewal projects, gleaming towers of conspicuous administration, gaudy commercial strips, quiet residential suburbs, the blank boxes and great parking lots of shopping malls, quaint heritage districts, industrial estates; then there are more modernist housing projects, more suburbs, another commercial strip, another industrial district, another post-modern townscape, another suburb. This is a strange sort of repetitive variety in which unity is achieved by contiguity and very little else. It is a real, spatial equivalent of TV programming in which, night after night, each precisely timed programme falls into a clear category such as comedy, drama, sports, current affairs and news, and has no relationship to what precedes or follows. It seems that modern life is filled with an easy acceptance of repetitive standardised discontinuities.

Urban places, except perhaps for new towns in their first few years, have probably always had a patchwork of landscapes. While some of these landscapes were clearly set apart by walls or by open spaces such as those around cathedrals, up until this century the majority of them were cheek-by-jowl and merged into one another. In the last 100 years what has happened is that the differences have been increasingly standardised and exaggerated, and then the entire cityscape has, as it were, been pulled apart along the seams. Like the expanding universe, everything seems simultaneously to have grown bigger and to have moved away from everything else. As buildings and artefacts have been made larger so the spaces and lines separating landscape districts have become more pronounced.

These changes and processes of segregation have, of course,

not been uniformly distributed everywhere. Landscapes dominated by pre-modernist forms and spaces still make up large areas of late twentieth-century cities. It is an indication of the success and attractiveness of these old street forms that they are so often vital and active, with crowded sidewalks, cafés and shops, as for example in Greenwich Village, or on Oxford Street in London and George Street in Sydney. These are neither heritage nor gentrified areas; instead their appearance has undergone continuous incremental change in which new styles have mostly been accommodated to the old forms. Modernity has nevertheless insinuated itself deeply into the fabric of these pre-modernist streetscapes, wrapping itself around the edges of buildings and leaving a distinctive deposit of plastic signs, plate-glass windows, pedestrianisation projects and corporate outlets like Marks and Spencer and Burger King.

It is in the cityscapes wholly made during the last 50 years that the manifestations of segregation are most apparent. Evidence of older landscapes has disappeared, except perhaps for fragments of street patterns and lot sizes, though in renewal projects new towns and suburban subdivisions, everything old, including roads and lots, has been systematically scraped away or buried under precisely counted tons of concrete and asphalt. It is here, in the suburban apartment complexes of Amsterdam, in the renewal projects and expressways of South Chicago, in Canberra, in the suburban subdivisions of Toronto and Los Angeles, in the new towns of Britain, that we can best see and judge the undiluted contributions of the twentieth century to urban landscape. The sight is frequently impressive, if only for the scale of what has been accomplished, and the efficiency of it all. Sadly, however, and perhaps this is precisely because of the scale and efficiency, this new landscape seems to promote few street activities or visual pleasures except where it turns to the old forms of post-modernism.

In the segregated landscapes of the modern city the lines of discontinuity often have a great significance. This is clearly so where new confronts old, for then the line between them constitutes a 'time edge', a boundary between two eras as well as between two landscapes. To emerge from a dingy old underground station and confront the massive concrete towers of the Barbican across the street, to drive from an area of victorian houses into a clean-sweep renewal project of apartments, or to encounter a turn-of-the-century building wedged between no-frills modernist offices, is to come upon a time edge (Figure 12.9). Such discontinuities play havoc with attempts at urban landscape criticism. Are we to judge the old in terms of the new, the new in terms of the old, or both in terms of themselves as though there are no time edges, or are we just to enjoy the variety? Enjoying the variety is hardly good enough. Abrupt edges of any sort reveal profound shifts in ideology, in social values, in the relative roles of development corporations, planning agencies and citizens. Their importance should not be underestimated, for even when they seem to reflect little more than changes in style they can mark deep social and cultural divisions. The campus of the Illinois Institute of Technology, with its low-rise, middle-class, Mies van der Rohe modernist design, is separated by a single street from the depressing and dangerous, scarcely less modernist high-rise public housing of South Chicago. To cross that street is to cross simultaneously an enormous social and economic gulf from affluence to poverty, from hope to despair, from a place where environmental control is taught to a place that is controlled. Even within the rectangular uniformities of modernist landscapes it is clear that some people are more equal than others. This is a street not easily crossed in either direction.

Marshall Berman writes in his book, *All that is solid melts into air* (1981), that the modern age is filled with paradoxes with its persistent tensions between the old and the new, its admiration of tradition and its desire for innovation, its love of order and desire for spontaneity, its overpowering bureaucracies and celebration of the individual. Modernity, he says (p. 15), and this is surely no less true for modern urban landscapes in particular, has to be grasped as 'a unity of disunity'.

This is certainly an important first step, and I believe that knowing something of how modern urban landscapes have emerged since 1880, with all the tensions between modernism and post-modernism, futurism and heritage, corporatisation and planning, is valuable preparation for making it. But this is hardly enough. Paradoxes beg to be resolved, and grasping urban landscapes as unities of disunity does not tell us either how to judge or how to change them. Are the modern urban landscapes of the late twentieth century good or bad? Are they successes or failures, or merely mediocre?

The easy and equivocating answer is that in some respects they are successful and in other respects they are failures. Obviously higher standards of health and convenience prevail

Figure 12.9: Social edges and time edges. The abrupt transition from the campus at the Illinois Institute of Technology in Chicago, designed by Mies van der Rohe, to some of the poorest public housing in America; this side and the other side of the street are worlds apart. A turn-of-the-century building on Swanston Street in Melbourne caught in a time warp

now than a century ago, and countless millions of people have been comfortably housed. It is certainly pleasant to be able to travel quickly and easily almost anywhere, for senior citizens to shop in the climate-controlled comfort of shopping malls, or to stroll around the reconditioned streets of Covent Garden. On the other hand, the hopes of Edward Bellamy and William Morris, of Ebenezer Howard, Raymond Unwin, Frank Lloyd Wright, Le Corbusier, and the members of the Bauhaus for city designs which would promote an egalitarian and just urban society, free from commercial exploitation and violence, and at ease with new technologies, simply have not been realised. The urban land-scape is littered with monuments to their failed dreams — neigh-bourhood units in which the elementary school has been closed, garden suburbs, congested expressways, Corbusian-style apart-ment projects surrounded by useless open space, isolated frag-ments of Usonia. Great planning ideas have repeatedly been rendered mediocre by standardisation, and magnificent paper projects, such as those of the Bauhaus for housing, have, on the ground, consistently turned out to be little better than huge prisons. To the extent that these hopes and aims could have been, but have not been, realised the modern urban landscape has to be judged a failure.

This is not all. Many of the built-forms and layouts of the modern city were conceived specifically as attempts to resolve social and economic injustices, essentially in the belief that good urban environments make good societies. These injustices per-sist, though they have changed their character and are not now as desperate as they were 100 years ago. In the late nine-teenth century the poor were all too apparent, their presence in the streets inspired Edward Bellamy and others to write about and to work for a different future. As a result of intensive plan-ning and welfare efforts the urban poor are now largely invisible, hidden in apartment projects like those of South Chicago, the no-frills modernist exteriors of which reveal little of the miseries within. Of course, the visible signs of deep inequalities can some-times be seen in the derelict landscapes of places like the South Bronx, but we are easily dissuaded from seeing these because they are out of the way and purported to be too dangerous to visit. In any case it is far easier to allow ourselves to be distracted by highly visible evidence of the apparently almost universal attainment of the good life in everything from flashy office towers to nice images and sentiments expressed on billboards or on benches in Malibu (Figure 12.10). Economic and social inequalities have not been eradicated as nineteenth-century reformers hoped, they have simply changed their form, and the paradox of poverty in an affluent society is resolved either by hiding it or by imagining it away. And it is not just poverty that has been made invisible.

Invisibility, whether of poverty, or toxic wastes, unemployment, the elderly or nuclear arsenals, is only one among many problems of modern urban landscapes. Consider the excessively rationalistic planning that lays out streets in the manner of a new by-law landscape, scarcely less uniform than its victorian predecessor; consider, too, the commodification of almost everything by corporations, the deceptions of imagineering, new roads which rip through the heart of old communities, new airports which obliterate old settlements, new buildings with windows which do not open and endlessly circulate mildly poisonous air, housing shortages, environmental degradation for quick profit-taking, and the terrorist bombings and street violence which condition the nervous ways in which we now have to experience many cities. For all the dramatic modifications that have been made to urban landscapes over the last 100 years I begin to suspect that the only fundamental social advances have been to do with sanitation. All the other changes — skyscrapers, renewal, suburban subdivisions, expressways, heritage districts — amount to little more than fantastic imagineering and spectacular window dressing.

This conclusion is too hard. It dismisses too easily the manifold achievements of twentieth-century city building. Modern urban landscapes are often visually spectacular, and generally reveal high levels of comfort, convenience and efficiency — they should be appreciated for these qualities. Of course they are also sometimes divisive, hard and deceptive, and for these qualities they should be condemned. Before we rush off to correct these problems it is well to remember, first, that a single landscape, for instance a public housing urban renewal development, can be both efficient and divisive, that is to say good and bad, at the same time; and then that many of these problems have arisen from the well-intentioned efforts of planners, architects and others who were probably no less concerned than us about the look of cities and the lives of their citizens, and some of whom had very high ideals indeed. Still at issue after more than a century of trying to resolve social problems by improved planning

Figure 12.10: Two concluding modern landscapes. A gutted building in the South Bronx in New York City is no less a part of the modern urban landscape than the sweet sentiments expressed on a bench in a shopping plaza in Malibu in California

and design is whether concerns about these problems can ever be focused in such a way that fundamental and enduring changes can take place. Such ambiguities, such persistent doubts, such mixtures of success and failure, show that a good urban environment, even if agreement could be reached on what this is, is not sufficient to create and sustain a just urban society. On a personal scale the circumstances of life have certainly been improved, but on the large scale it turns out that modern urban landscapes, like all landscapes, are reflections rather than causes of the human condition.

Bibliography

Adams, H. (1938) *Letters of Henry Adams 1892-1918*, Vol. 2, Houghton Mifflin, Boston

Adams, T, (1932) *Recent advances in town planning*, J. & A. Churchill, London

—— (1931) *The building of a city: the regional plan of New York and its environs*, Vol. 2, Port of New York Authority

Alexander, C. (1975) *The Oregon experiment*, Oxford University Press, New York

American Institute of Planners (1953) 'Defense considerations in American planning', *Bulletin of the Atomic Scientists*, 9, 268

Andrew, E. (1981) *Closing the iron cage: the scientific management of work and leisure*, Black Rose, Montreal

Andrews, W. (1964) *Architecture, ambition and Americans*, Free Press, New York

Appleyard, D. (ed.) (1979) *The conservation of European cities*, MIT Press, Cambridge, MA

Architectural Record (1925) 58, 373

Atwan, R., McQuade, D. and Wright, J.W. (1979) *Edsels, Luckies and Frigidaires: advertising the American way*, Delta Publishing, New York

Banham, R. (1960) *Theory and design in the first machine age*, Architectural Press, London

—— (1963) 'Brutalism', in G. Hatie (ed.), *Encyclopaedia of modern architecture*, Thames and Hudson, London, 61-4

—— (1976) *Megastructures: urban futures and the recent past*, Harper and Row, New York

—— (1984) *The architecture of the well-tempered environment*, University of Chicago Press

Barnett, J. (1982) *An introduction to urban design*, Harper and Row, New York

Bauer, C. (1934) *Modern Housing*, Houghton Mifflin, New York

Belasco, W.J. (1979) *Americans on the road*, MIT Press, Cambridge

Bell, D. (1973) *The coming of post-industrial society*, Basic Books, New York

Bellamy, E. (1888) *Looking backward: 2000-1887*, New American Library, New York (1960)

—— (1897) *Equality*, Appleton and Company, New York

Benevolo, L. (1960) *History of modern architecture*, Vol. 2, The Modern Movement, MIT Press, Cambridge, MA

—— (1981) *The history of the city*, MIT Press, Cambridge, MA

Berman, M. (1981) *All that is solid melts into air*, Simon and Schuster, New York

Blake, P. (1960) *Frank Lloyd Wright*, Penguin, Baltimore

—— (1964) *God's own junkyard*, Holt Rinehart and Winston, New York

—— (1977) *Form follows fiasco: why modern architecture hasn't worked*, Little Brown, Boston

—— (1982) 'The end of cities', in L. Taylor (ed.), *Cities: the forces that shape them*, Rizzoli, New York

Boorstin, D. (1973) *The Americans: the democratic experience*, Random House, New York

Boyd, J.T. (1920) 'The New York zoning resolution and its influence on design, *Architectural Record*, 48, 193-217

Boyer, M.C. (1983) *Dreaming the rational city*, MIT Press, Cambridge, MA

Brett, L. (1971) *Architecture in a crowded world*, Schocken Books, New York

Brolin, B. (1976) *The failure of modern architecture*, Van Nostrand Reinhold, New York

—— (1980) *Architecture in context: fitting new buildings with old*, Van Nostrand Reinhold, New York

Buchanan, C. (1963) *Traffic in towns*, Penguin, Harmondsworth

Cherry, G.E. (1974) *The evolution of British town planning*, John Wiley, New York

Christensen, T. (1982) 'A sort of victory: Covent Garden revisited', *Landscape*, 26, No. 2, 21-8

Clawson, M. and Hall, P. (1973) *Planning and urban growth*, Johns Hopkins University Press, Baltimore

Clunn, H. (1932) *The face of London*, Simpkin Marshall, London

Coleman, A. (1985) *Utopia on trial*, Hilary Shipman, London

Collier, R.W. (1974) *Contemporary cathedrals: large scale developments in Canadian cities*, Harvest House, Montreal

Corbett, M.N. (1981) *A better place to live: new designs for tomorrow's communities*, Rodale Press, Emmaus, PA

Creese, W.L. (ed.) (1967) *The legacy of Raymond Unwin*, MIT Press, Cambridge, MA

Crosby, T. (1973) *How to play the environment game*, Penguin, Harmondsworth

Cullen, G. (1971) *The concise townscape*, Architectural Press, London

David, A.C. (1910) 'The new architecture', *Architectural Record*, 28, 389-403

De Chiara, J. and Koppelman, L. (1975) *Urban design criteria*, Van Nostrand Reinhold, New York

Drexler, A. (1979) *Transformations in modern architecture*, Museum of Modern Art, New York

Edwards, A.M. (1981) *The design of suburbia*, Pembridge Press, London

Ehrlich, P., Ehrlich, A. and Holdren, J. (1973) *Human ecology*, W.H. Freeman, San Francisco

Evenson, N. (1979) *Paris: century of change*, Yale University Press, New Haven

Essex County Council (1973) *A design guide for residential areas*, County Council of Essex

Fishman, R. (1977) *Urban utopias in the twentieth century*, Basic Books, New York

Fitch, M. (1947) *American building 1: the historical forces that shaped it*, Houghton Mifflin, Boston

—— (1961) *Architecture and the esthetics of plenty*, Columbia University Press

Frampton, K. (1980) *Modern architecture: a critical history*, Thames and Hudson, London

Francis, M. (1983) 'Community design', *Journal of Architectural Education*, 37, 1, 14-19

Freysinnet, E. (1934) in *Architectural Record*, 75, 41

Fryer, W.J. (1891) 'Skeleton construction', *Architectural Record*, 1, 228-35

Fussell, P. (1975) *The Great War and modern memory*, Oxford University Press, New York

Futterman, R.A. (1961) *The future of our cities*, Doubleday, New York

Galbraith, J.K. (1968) *The new industrial state*, Signet Books, New York

Gans, H. (1967) *The Levittowners*, Pantheon Books, New York

—— (1968) *People and plans*, Basic Books, New York

Goldberger, P. (1983) *On the rise: architecture and design in a postmodern age*, Penguin, New York

Goodman, R. (1971) *After the planners*, Simon and Schuster, New York

Gropius, W. (1965) *The new architecture and the Bauhaus*, MIT Press, Cambridge, MA

Gruen, V. (1964) *The heart of our cities*, Simon and Schuster, New York

Guth, A.G. (1926) 'The automobile service station', *Architectural Forum*, 45, 35-56

Handlin, D. (1979) *The American home: 1815-1915*, Little Brown, Boston

Harvey, T. (1984) 'Federal housing policy and the design of suburbs', paper presented to Association of American Geographers, Annual Conference, Washington, DC

Hayden, D. (1981) *The grand domestic revolution*, MIT Press, Cambridge, MA

Hester, R. (1985) 'Subconscious landscapes of the heart', *Places*, 2, No. 3, 10-22

Hilbersheimer, L. (1964) *Contemporary architecture*, Paul Theobold, Chicago

Hill, G. (1895) 'Some practical limiting conditions in the design of the modern office building', *Architectural Record*, 2, 446-68

Hines, T.S. (1974) *Burnham of Chicago: architect and planner*, Chicago University Press, Chicago

Hitchcock, H-R. (1948) *Painting toward architecture*, Duell, Sloan and Pearce, New York

—— (1958) *Architecture: nineteenth and twentieth centuries*, Penguin, Harmondsworth

—— and Johnson, P. (1932) *The international style*, W.W. Norton, New York (1966)

Hoskins, W.G. (1955) *The making of the English landscape*, Penguin, Harmondsworth

Howard, E. (1902) *Garden cities of tomorrow*, Faber and Faber, London (1965)

Hughes, R. (1980) *The shock of the new*, BBC Publications, London

Huxtable, A.L. (1972) *Will they ever finish Bruckner Boulevard?*, Collier Books, New York

Jacobs. A. (1985) *Looking at cities*, Harvard University Press, Cambridge, MA

Jacobs, J. (1961) *The death and life of great American cities*, Vintage Books, New York

James, H. (1907) *The American scene*, Charles Scribners Sons, New York (1946)

Jencks, C. (1977) *The language of post-modern architecture*, Rizzoli, New York

Johnson-Marshall, P. (1966) *Rebuilding cities*, Aldine Publishing, Chicago

Kennedy, W. (1983) 'Everything everybody ever wanted', *Atlantic Monthly*, May, 77-88

Langdon, P. (1985) 'Burgers! Shakes!', *Atlantic Monthly*, December, 75-89

Le Corbusier (1923) *Vers une architecture*, Arthaud, Paris (1977)

—— (1929) *The city of tomorrow and its planning*, Architectural Press, London (1971, original French edition 1924)

Leinberger, C.B. and Lockwood, C. (1986) 'How business is reshaping America', *Atlantic Monthly*, October, 43-52

Lessard, S. (1976) 'Reflections: the suburban landscape — Oyster Bay, Long Island', *The New Yorker*, 11 October 1976, 44-79

Lifton, R.J. (1979) *The broken connection: on death and the continuity of life*, Simon and Schuster, New York

Lonberg-Holm, K. (1930) 'The gasoline filling and service station', *Architectural Record*, 67, 563-84

Lorimer, J. (1978) *The developers*, J. Lorimer, Toronto

Lowenthal, D. and Binney, M. (eds) (1981) *Our past before us?*, Temple Smith, London

Lynch, K. (1971) *Site planning*, MIT Press, Cambridge, MA

Lynes, R. (1949) *The tastemakers*, Harper and Brothers, New York

MacDonald, K. (1985) 'The commercial strip: from main street to television road', *Landscape*, 28, No. 2, 12-19

McKelvey, N. (1968) *The emergence of metropolitan America: 1915-61*, Rutgers University Press, New Jersey

Marinetti, F.T. (1909) 'The founding and manifesto of futurism', in R.W. Flint (ed.), *Marinetti: selected writings*, Farrar, Straus and Giroux, New York (1971)

Marriott, O. (1967) *The property boom*, Pan Books, London

Marsh, B.C. (1909) *An introduction to city planning*, Arno Press, New York (1974)

Mayer, H.M. and Wade, R.C. (1969) *Chicago: growth of metropolis*, University of Chicago Press, Chicago

Mearns, A. (1883) *The bitter cry of outcast London*, Leicester University Press, Leicester (1970)

Meyrowitz, J. (1985) *No sense of place*, Oxford University Press, New York

Morris, M. (1966) *William Morris — artist, socialist, writer*, Russell and Russell, New York

Morris, W. (1890) *News from nowhere*, Routledge and Kegan Paul, London

—— (1962) *Selected writings* (ed. Asa Briggs), Penguin, Harmondsworth

Mumford, L. (ed.) (1952) *Roots of contemporary American architecture*, Reinhold, New York

—— (1970) *The myth of the machine*, Vol. Two: The Pentagon of Power, Harcourt, Brace, Jovanovich, New York

Nairn, I. and Cullen, G. (1955) *Outrage*, Architectural Press, London (1959)

Oliver, P. *et al.* (1981) *Dunroamin: the suburban semi and its enemies*, Barrie and Jenkins, London

Osborn, F.J. and Whittick, A. (1977) *New towns*, Leonard Hill, London

Papageorgiou, A. (1971) *Continuity and change: preservation in city planning*, Praeger, New York

Perry, C. (1929) 'The neighborhood unit', in Neighborhood and community planning, *The Regional Plan of New York and its environs*, Vol. 3, Port of New York Authority

Pevsner, N. (1968) *The sources of modern architecture*, Praeger, New York

Picturesque World's Fair (1894) H.B. Closkey, Chicago

Pond, I.K. (1921) 'Zoning and the architecture of high buildings', *Architectural Forum*, 35, 131-4

Popenoe, D. (1977) *The suburban environment*, University of Chicago Press, Chicago

Rae, J.B. (1971) *The road and the car in American life*, MIT, Cambridge, MA

Raskin, E. (1974) *Architecture and people*, Prentice Hall, New York

Ravetz, A. (1980) *Remaking cities*, Croom Helm, London

Reynolds, D.M. (1984) *The architecture of New York City*, Macmillan, New York

Rimbert, S. (1973) *Les paysages urbaines*, Armand Colin, Paris

Riseboro, B. (1979) *The story of western architecture*, MIT, Cambridge, MA

Ruskin, J. (1865) *Crown of wild olive*, Mershon Company, New York

San Francisco (1971) *The urban design plan for the comprehensive plan of San Francisco*, Department of City Planning, San Francisco

Schopfer, J. (1902) 'A new method of cement construction', *Architectural Record*, 12, 271-80

Schuyler, M. (1894) 'Last words about the World's Fair', in *American architecture and other writings*, W.H. Jordan and R. Coe (eds), Belknap Press of Harvard University Press, Cambridge, MA (1964)

Schwilgin, F. (1973) *Town planning guidelines*, Department of Public Works, Ottawa

Scott, G. (1914) *The architecture of humanism*, University Paperbacks, London (1961)

Scott, M. (1969) *American city planning since 1890*, University of California, Berkeley

Sharp, T. (1940) *Town planning*, Penguin, Harmondsworth

—— (1948) *Oxford replanned*, Architectural Press, London

Singer, C. *et al.* (eds) (1958) *A history of technology*, Vol. V: The Late Nineteenth Century, Oxford University Press, Oxford

Sitte, C. (1889) *The art of building cities*, Hyperion Press, Westport, CN (1949)

Sommer, R. (1974) *Tight spaces: hard architecture and how to humanize it*, Prentice Hall, New York

Stein, C.S. (1958) *Toward new towns for America*, Liverpool University Press

Stilgoe, J. (1982) *Common landscapes of America: 1580-1845*, Yale University Press, New Haven

Strong, A.L. (1971) *Planned urban environments: Sweden, Finland, Israel, The Netherlands, France*, Johns Hopkins Press, Baltimore

Sullivan, L. (1896) 'The tall office building artistically considered', in *Kindergarten chats and other writings*, George Wittenborn, New York (1947)

—— (1923) 'The Chicago Tribune competition', *Architectural Record*, 53, 157

Sutcliffe, A. (1981) *Towards the planned city*, Basil Blackwell, Oxford

Swan, H.T. (1921) 'Making the New York zoning ordinance better', *Architectural Forum*, 35, 125-30

Taylor, F.W. (1911) *Scientific management*, Greenwood Press, Westport, CO (1947)

Towndrow, F. (1933) *Architecture in the balance*, Chatto and Windus, London

Tunnard, C. and Pushkarev, B. (1963) *Man-made America: chaos or control*, Yale University Press, New Haven, CO

Unwin, R. (1909) *Town planning in practice: an introduction to the art of designing cities and suburbs*, T. Fisher Unwin, London

Updike, J. (1965) *Assorted prose*, Alfred Knopf, New York

Urban Land Institute (various editions from 1947 to 1968) *Community builder's handbook*, Urban Land Institute, Washington, DC

—— (1977) *Shopping center development handbook*, Urban Land Institute, Washington, DC

Veblen, T. (1899) *The theory of the leisure class*, Modern Library, New York (1934)

—— (1923) *Absentee ownership and business enterprise in recent times*, B.W. Huesch, New York

Venturi, R. (1966) *Complexity and contradiction in architecture*, Museum of Modern Art, New York

—— (1972) *Learning from Las Vegas*, MIT Press, Cambridge, MA

Warner, S.B. (1972) *The urban wilderness: a history of the American city*, Harper and Row, New York

Wattel, H.L. (1958) 'Levittown: a suburban community', in Dobriner, W.M. (ed.), *The suburban community*, G.P. Putnam's Sons, New York

Whitehouse, R. (1980) *A London album*, Secker and Warburg, London

Whittick, A. (ed.) (1974) *Encyclopedia of urban planning*, Robert E. Krieger, Huntington, NY

—— (1950) *European architecture in the twentieth century*, Crosby Lockwood, London

Whyte, W.H. (1982) *The social life of small urban spaces*, Conservation Foundation, Washington, DC

Williams, J.D. (1923) 'Advertising signs removed from state highways', *The American city*, 28, 5, 484-5

Williams-Ellis, C, (1928) *England and the octopus*, Geoffrey Bles, London

Wolfe, T. (1981) *From Bauhaus to our house*, Farrar, Straus, Giroux, New York

Wright, F.L. (1901) 'The art and craft of the machine', in L. Mumford (ed.), *The roots of contemporary American architecture*, Reinhold, New York (1952)

—— (1927) 'In the cause of architecture', *Architectural Record*, 61, 394-6

—— (1935) 'Broadacre City: a new community plan', *Architectural Record*, 77, 243-54

—— (1943) *An autobiography*, Duell, Sloan and Pearce, New York

—— (1945) *When democracy builds*, University of Chicago Press, Chicago

—— (1958) *The living city*, Horizon Press, New York

Wright, P.B. (1910) 'Additions to Chicago's skyline', *Architectural Record*, 28, 15-24

Index

Index

Index

Printed and bound by CPI Group (UK) Ltd, Croydon, CR0 4YY

22/10/2024

01777623-0002